Glamour and Geology

E. Allen Driggers

Glamour and Geology

Women in Petroleum Geology and Popular Culture

E. Allen Driggers
Tennessee Technological University
Cookeville, TN, USA

ISBN 978-3-031-64524-2   ISBN 978-3-031-64525-9 (eBook)
https://doi.org/10.1007/978-3-031-64525-9

This Springer imprint is published by the registered company Springer Nature Switzerland AG
The registered company address is: Gewerbestrasse 11, 6330 Cham, Switzerland

This book is dedicated wholeheartedly to
the most accomplished woman I know:
Dr. Laura Elizabeth Smith

Foreword: "Cutting Down the Nets"

Readers: it might be a strange start to a book about geology to start with a basketball analogy, but unconventional seems to fit this book just fine. Basketball seems to summarize the mood and circumstances of its creation. In 1983, a college basketball team from Raleigh, North Carolina, had an unconventional run through the National Collegiate Athletic Association (NCAA) men's basketball tournament and many North Carolinians think of them as the "Cinderella Team." The meaning of Cinderella team is a sports metaphor for a team that seemed to win, despite their lack of pedigree, traditional team structure, and seemed to win against more competitive opponents. The team from North Carolina State University is most well known for their coach, a charismatic New Yorker and Italian-American named James Thomas Anthony Valvano (or Jimmy V), who seemed to stir up the team. Valvano was an English major who used quotations from literature to inspire the team. Valvano is also remembered for his work in raising funds for cancer researcher through his foundation and his signature phrase "Never give up." Valvano suffered from cancer himself. He died in 1993, but his mythological status was already transfixed in my mind growing up in the Raleigh-Durham area, as my parents, neither of them graduates of the North Carolina State University, but two persons who became enraptured in the spirit of 83.

When you try to explain major popular culture events that your spouse was not familiar with, given her upbringing in Arkansas, far away from the basketball courts of the Atlantic Coast Conference, you sometimes need moving images and film that explains the significance of something like the 1983 championship. Luckily, the new *30 For 30* film covering the folklore and events of the 1983 was available, entitled *Survive and Advance*, and Laura Elizabeth Smith, my fellow academic and wife, was able to make the connections and see things through my eyes, which is crucial in any relationship.

During 2022, my wife was going on the job market and finishing her Ph.D. Both of these events span the creation of the book, including the beginnings of our dating history. When anyone is either writing a doctoral thesis of being on the job market, or even doing both at the same time, there can be moments where your dreams seem impossible. The same can be said about starting your second book, without an advisor and balancing the responsibilities of a full-time academic job. Though our future together was certain, our working plans were painfully uncertain. But, she too is charismatic, and recalled the scene from the documentary where at the beginning of the 1982–1983 season, Valvano gathered the team together, at the first practice, and got them thinking about winning the national championship as a goal, which seemed as impossible as two academic jobs in the liberal arts. To get the team focused on realizing that goal, Valvano had them practice everything, including cutting down the nets, which was done at the end of winning a championship.

In that same spirit, Laura bought her doctoral gown on the first day she started revising her dissertation and I, too, started writing the forward, preface, and acknowledgments to this book. And though things are way different in 2024 than in 1983, with Caitlin Clark being one of most well-known basketball players, and the game going through more diverse and hopefully inclusive changes, the attitude of hope remains. As I write this, both the North Carolina State University men and women's basketball teams are slated to play in the final four rounds of the NCAA tournament, and they have the opportunity to win the national championship with two more games. Whether you care about sports or not, the spirit of reaching for something that seems so impossible and out of reach never seems to go out of fashion, and perhaps a crucial part to realizing your goals is to practice the whole experience, including how to celebrate them.

Though my goals were comparatively smaller and I had significant resources behind me to achieve them (see the acknowledgments) some of my smallest goals seemed on that same, impossible scale. But writing this book has showed me that with community, support, and a little charisma, you can be pushed into some amazing experiences in life. Women featured in the book are quite frankly amazing, inspiring, and successful, whether they were pushing against the large mechanics of sexism or simply going in and doing their job everyday despite a toxic culture.

Preface: GTT

In the nineteenth century, when persons wanted to escape debts, run away, change their lives for the better, or seek adventure, they often fled states like South Carolina and left a simple note explaining their whereabouts: or left just enough information not to be found. The note often read "GTT." The abbreviation stood for "Gone to Texas" (Campbell 2017). Writing this book meant all of those things to me and more. One day, at the start of one amazing research trip I left the Facebook status "GTT." Going to Texas changed my life for the better and it was truly one amazing trip that lead to many others.

The narrative arc of my life during that first Texas research trip was trending into an amazing change, and I was moving out of some rather dark rain storms. If you would like more back story, please consult the preface in E. Allen Driggers, *Early Nineteenth Century Chemistry and the Analysis of Urinary Stones* (Driggers 2023). The first time I traveled out to Lubbock, Texas, I viewed it as a place of mystery and excitement. I was going West! I had just met my soon to be wife, Laura Elizabeth Smith, but she had some understandable trepidation about making a commitment with someone who was nine hours and almost six hundred miles away. I saw her and the West as my destiny and embraced the experience, traveling to the archives all day long and talking to her on the phone late into the evening, almost missing my flight from Nashville to Dallas.

Traveling to Lubbock was intimidating, as the flight required multiple plane embarkations and though I had traveled to Europe, I had never gone West of Oxford, Mississippi. However, my undergraduate mentor at North Carolina State University, the highly accomplished plant pathologist Larry Blanton, formerly of Texas Tech University, introduced me to his social network of quite frankly, amazing Texans. I would like to thank Cal and Melanie Barnes, Julie S. Isom, Celeste Yoshinobu, Andy Wilkinson, and many other kind members of the Texas Tech Community, go Red Raiders! I will be a fan of the Red Raiders not because Texas Tech funded any travel, but people who are associated with Lubbock are so damn kind.

In 2019, I was able to return with Laura, convincing her to ride all the way from Tennessee without air conditioning. Both Dr. Smith and I would like to thank Lewis Held of the biology department. After a random run in, where we gave us "Karma

Dollars" to honor our recent engagement, we count him as an awesome friend! Held was encouraging of both book projects.

With the most enjoyable of projects, often historians come to these projects by complete accident. I had developed a conference paper for the Geological Society of America meeting in Indianapolis from the remnants of the personal papers of the geologist Augusta Hasslock-Kemp that were held in Tennessee. I thought it would be my one foray into the history of geology, but a publisher from Springer persuaded me to publish a book on the geologist. I took an amazing trip to Lubbock, Texas, and performed research at the Southwestern Collection. There I found a fascinating newspaper article profiling Kemp and other female geologists working in the oil fields of America during World War II. As a naive, and passionate (tenure track) researcher, I thought I could simply will the book into existence, as I was confident biographical information about these women had to be out there. I got my contract and I was off to the races. I anticipated multiple trips to Texas from Tennessee and that the book would practically write itself. I could not have been more incorrect, but in the best of ways. The COVID pandemic happened and changed the United States in ways that I could not have anticipated, and forced me to re-think through a research project using new to me resources. Newspapers became my lifeline, as well as digitized books. However, the meaning of all of it was unclear, and I wondered how was I ever to finish what felt was a monumental two hundred and fifty pages. I wrote an occasional chapter here and there, but could never get any momentum going toward fully completing the book.

I looked for ways to be re-energized about the project, as the pandemic and online teaching proved to be challenging. Laura Smith and I's daughter, Millie, also started online kindergarten during the pandemic. Online kindergarten challenged our abilities as teachers at work on our own classes but also making sure Millie had enough support for school. I sat beside her each morning at Laura's great grandmother's dining room table and tried to make sense of the research project, sitting in so many pieces, much like a quilt, and looking for an assembling and a coherent strategy for completion. But mostly I simply started reading newspapers and seeing what caught my eye. I also simply prayed that I would be led, in a great leap of faith, to a coherent and interesting story. Sometimes you simply have to read through newspapers, hope, and write. Eventually, the narrative appeared.

Hope was supplied for this project from the confidence that came from the publication of my first book with Springer, *Early Nineteenth Century Chemistry and the Analysis of Urinary Stones*. The publication of a book about urinary stones was a long and challenging process, where many academic publishers at University presses never saw the market for the book. A lot of rejections and near publications eventually gave way to a yes, and I felt that I could do anything after that book came out on July 14, 2023. Laura Smith would not let me give up on that book, as she and Kent Dollar talked me in to continuing to try and find a publisher for the book. When the first book came out, she implied that I could now focus and finish the second book: she was correct and I greatly underestimated my determination, stubbornness, and abilities of a historian. I was able to report to her throughout the summer all the progress I was making by simply going into my office a couple of days a week throughout the summer and the book simply wrote itself. I am lucky that Laura Smith believes in me all the times I do not, and I hope to have two people (including myself) supporting my next book project.[1] Anything is possible with love.

Cookeville, TN, USA E. Allen Driggers

References

Campbell RB (2017) Gone to Texas: a history of the Lone Star State. Oxford University Press, New York

Driggers EA (2023) Early nineteenth century chemistry and the analysis of urinary stones. Springer, New York

[1] Jude 1:2 Mercy unto you, and peace, and love, be multiplied.

Acknowledgments

This book was completed though assistance and distraction from Millie Driggers; she's doing a fantastic job growing up! My mother Jean Driggers, inspired my love of history at an early age, and she and my father took me to many historical sites that cemented that interest. Thank you, Sara Smith and Grant Smith, for looking after Millie and the many gifts of used furniture. Thank you to Susan Laningham for the Tuesdays free to work and play racquetball. I hope that Millie Driggers reads this book out loud as many times as she read first book: that was really touching! Thank you to my father, Ed Driggers, resident of the world hereafter, for all of his encouragement.

I received financial support from Tennessee Technological University and the faculty research fund. I have some very good colleagues at Tech that were helpful with the completion of this book: Dr. Kent Dollar, Dr. Jeff Roberts, Dr. Alan Mills, Dr. Troy Smith, and Dr. Laura Elizabeth Smith. In many cases I wanted to end the book project, as I felt overwhelmed or existential and they did not let me give up, much in the spirits of the words of James Valvano. Thank you David Hales for help with images and great racquetball games. Thank you to Amy Foster for copying services and logistical support with getting all those books to and from the library. Thank you, Dr. Ronald Brashear, for a kind word when I needed it. Appreciation to Dr. Ben Gross for his dependability and encouragement. I am greatly indebted to Dr. Laurence Hare at the University of Arkansas, as he has provided crucial support and mentoring. I would also like to thank Dr. Adelheid Voskuhl at the University of Pennsylvania for acting in the true kind-person-scholar vibe of her friend and my mentor, Dr. Ann Johnson. Thank you to Andy Wilkinson at Texas Tech for hospitality when I visited Lubbock. Dr. Lewis Held, thank you for the Karma dollars, Laura and I are investing them in kind persons. Many people at Texas Tech helped and again, I will thank you again: Cal and Melanie Barnes, Julie S. Isom, and Celeste Yoshinobu. Thank you to the Southwest Collection at Texas Tech for image permissions.

I would like to thank my editor Aaron Schiller for continuing my project, as the original editor was called to do different work. He was an excellent advocate, encouraged me, and was patient in my anxious and early stages of the project.

Schiller worked with me through the pandemic and was there for me anytime I needed him; excellent work! I would like to thank Dr. Shirley Laird and Philip Davis for excellent dinners at Nick's, they raised both Laura and I's spirits! He would also like to thank Scott Moss and Gail Moss for their friendship and culinary adventures! Thank you to Troy and Robin Smith for excellent breakfast get togethers in Sparta.

Librarians at Tennessee Tech and other research institutions were instrumental in getting me resources of helping me track down leads. Holly Mills helped me with sources from interlibrary loan and answered a half billion of my desperate emails. Ashley Augustyniak McDermott at Science History Institute helped me with gathering rare primary sources. Beth Lander, as always, was helpful with encouragement and sources. Dr. David Dangerfield at the University of South Carolina-Salkehatchie helped me find rare geology publications and kept me from quitting the project. I am also incredibly thankful to the staff at the Dolph Briscoe Center for American History for sending me some scanned material during the middle of the COVID-19 pandemic. Thank you to Erin A. Harbour for her assistance in tracking down and double-checking sources. I also prayed for intercession and guidance to Dr. Ann Johnson, formerly of the University of South Carolina, now philosopher of the world hereafter. Again, Dr. Larry Blanton was incredibly helpful establishing my Lubbock connection. I would also like to thank Jim Flis for emailing me out of the blue about Augusta Kemp and getting me re-engaged with my book project. Thank you to Tom Vance for sharing his files and materials about Kemp, I sincerely appreciate it! Again, thank you Robbie Gries for being generous in time and spirit, and helping me with leads and sources. You were kind to talk to me, mail me sources, and encourage me to add another book to the needed histories of women in petroleum geology. Thank you to the Bitgood, Decker, and McCoy families for sharing access to documents and photographs of family members. Again, thank you Robbie Gries for access to transcribed and painstakingly gathered documents and photographs you used for your own book; that was a big help. I hope this book adds to the rich narrative of women in geology.

The largest source of confidence came from my wife, Dr. Laura Elizabeth Smith. She toiled with her own dissertation, teaching load, and then her job market sojourn. She truly took a piece of coal and made it into a diamond, and made it through trials and tribulations that just a single challenge would ruin most people, much less a series of them. She's the best scholar, mother, decorator, craftswoman, and just good buddy that I have ever known and will know. She never let me give up on this project and would gladly defend my honor. Laura truly understands the value of geological prospecting, as it takes only one oil well to truly produce a fortune. And in that respect, I am the luckiest prospector out there. Now, to the delight of many people that I thanked in these acknowledgments, I am off to clean up my office and return all my books to the library, and get prepared for the next project!

Contents

List of Figures

List of Tables

About the Author

E. Allen Driggers is an Associate Professor of history at Tennessee Technological University. The last nine years or so have been a wild ride, but have produced the best of circumstances, such as his wife Dr. Laura Elizabeth Smith. He has published one book about the history and chemistry of urinary stones at the turn of the nineteenth century entitled *Early Nineteenth Century Chemistry and the Analysis of Urinary Stones*. He has published scholarly articles in *Critical Philosophies of Race*, *Water History*, *The Journal of Medical Biography*, *The South Carolina Historical Magazine,* and *The History of Psychiatry*. With his co-author Laura Elizabeth Smith, he has published articles on the nature of race and the history of science in publications like *Earth Science History, Journal of Sleep Research*, and *Baylor University Medical Center Proceedings*. He is an affiliated researcher with The Whiteside Museum of Natural History in Seymore, Texas. Driggers is excited for his next trip to explore Texas! He is working on two new book length projects on the history of medicine and scientific societies in India and a history of the chemistry of inflammation. He lives with his wife Laura, their daughter Millie, two dogs, and one cat in the former capital of the state of Tennessee: Sparta.

Chapter 1
Imagination and History: Glamour and Geology

1.1 The Glamour of Geology

Women geologists were portrayed as *glamourous* during World War II and after, though the notion of glamour changed over time. The job more or less made women locally famous but put a lot of pressure on them to conform to societal standards and expectations. Women were portrayed as stylish, glamourous, heroic, intelligent, but humble and grounded. The female geologist was portrayed as adventurous and romantic. Women geologists were praised for their good looks and movie star style, and their experiences working in the petroleum field were promoted in newspapers that circulated around the United States. The acknowledgement that there were not many female geologists working in fields like petroleum exploration became attached to narratives about these working women. But there was a message of opportunity for women to enter this exciting field and to be exceptional. The door to access education and jobs in petroleum geology seemed to be open to any woman, as advertised in print media like newspapers. Though in actuality, these doors were only open to certain groups of women and often excluded African American working women (Morris 2021). The oil industry, in the twentieth century, advertised that there were not only opportunities for working female geologists but women who could advertise, translate articles in foreign languages, write promotional stories about the oil industry for the newspaper, and work in supporting roles like childcare and oil transportation (Ponton 2019).

Petroleum geology was a dirty job. But newspapers advertised the practical glamour in the everyday workings of women in geology and petroleum. Both the industry and sometimes the women themselves fueled and created the image and narrative of a woman working in the glamourous field of petroleum. However, women's experiences working in geology and the petroleum field were not perfect and not as romantic as portrayed in newspaper ink. Women had to adhere to a

E. A. Driggers, *Glamour and Geology*,
https://doi.org/10.1007/978-3-031-64525-9_1

hyper-feminized image of women working in petroleum and geology to justify their place in their jobs and science (Meyer 1996; Honey 1984; Williams 2021).

Newspapers continually ran stories from the late nineteenth century into the 1980s about the daring and hyper-feminized working woman geologist. They were often portrayed as above-average woman, doing amazing work in a mostly male-dominated field. These women were portrayed and described by the standards of physical beauty at the time, and some even appeared in the pages of men's magazines of the 1970s, like *Playboy* (Pitzulo 2011; Dyhouse 2010). Scholars of the postwar period noted in similar studies that women had pressure to adhere to the feminization of jobs that became open to women as the need for labor increased during World War II (Milkman 1987; Meyerowitz 1994). In this study, I argue that women also felt pressure to be portrayed in these *glamourized* terms and hyper-feminized language to advance and increase their participation in petroleum geology, roughly from the Victorian era to the 1980s (Gorham 1982; Stone and Sanders 2021).

The pressures that these women experienced must have been extreme and unique from those of their male counterparts. Women in geology, who appeared in newspapers, seem to not overtly fight against the image portrayed in the press until the 1970s. The outside social changes and larger social movements likely inspired some women to fight against the *imagined* image of the female geologist working in petroleum (Cobble 2015). There was evidence that women pushed back, including objecting to imposed clothing and fashion expectations at work or trying rewrite the image of the female geologist, such as in the case of Kathryn Sullivan, an astronaut in the NASA space shuttle program (Foster 2011). During the second women's rights movement in the 1970s, women tried to assert themselves into the *imagination* of what a female geologist would be perceived by the public and tried to have more agency and input into what the *image* of the female geologist would be from the 1970s onward.

I argue that aspects of the romantic *image* of the female petroleum geologist remained in perceptions of the *Oil Woman* and in toys for young women, such as *Barbie*, well into modern memory. Women now have much more control over the image of the female scientist and far more availability to advocate and craft their own trajectories in science, but *aspects* of romantic images and ideas are still evoked in narratives that focus on the "trailblazing" female scientist and in the female scientists that were fighting against the inherent sexism, classism, racism, and other inequities of science. Often, narratives forget the everyday woman working in science who goes to work every day or does not consider herself a radical fighting against the larger patriarchal system or resists the system in more subtle ways. Who will tell those stories?

The challenge of a lot of histories written about women is that, though they explain a lot of bravery and struggle of women in science, there are too few biographies of average, working women in science. Women fighting inequality in science are important and should be taught in classrooms at all levels, but it needs to be balanced with the stories of everyday women who made their mark, even though those writings might not exist or be preserved in anthologies of women's lives, such as on Wikipedia or in print. The narratives of women who found themselves confused or ambivalent about the larger social changes going on in society or science also matter as they flesh out the total female experience in geology. Acknowledging the difficulties of women's lives in science at all levels shows the sufferings of inequalities. These narratives are also important in exploring and witnessing struggle, and they are important in historicizing science.

If people perceive women who study science as always having to be exceptional, the pressure to excel might be a deterrent in and of itself. The pressures to be first or the most exceptional have been examined by psychologists and found to be detrimental to the undergraduate experience (Ebony 2020; Arredondo et al. 2022). Men who enter science, or even history, lack the pressures to make lasting and forever contributions to their field and have the freedom to write books that will likely not win awards or gather great acclaim (Stewart-Williams et al. 2021). I include myself as I strive to write books to be the best that I can, but I am not under the pressure of writing a National Book Award-winning book. I am free to make my career in any way that I please, and I think writing more histories of everyday working women in science and the humanities would be helpful to women aspiring to pursue higher education in all its forms (Driggers 2023). I would argue that *everyday exceptionalism*, or women who were working and free to pursue their own interests and their own research questions, would encourage more equity in the sciences, in addition to salary increases, better access to facilities and programs, as well as better representation.

This book, in many ways, is an *exploration* of women the reader is likely not familiar with but who were considered *newsworthy* in their own times. Newspapers make up a lot of the resources that tell the narrative of the book because they illustrate the *imagination* and *glamourous* images of women in geology, often involved with the petroleum industry. Newspapers were consumed by many different parts of American, Canadian, and British society, though this study focuses mostly on the United States (Wallace 2005). The common consumption of newspapers by many people in the past gives historians a pulse of intellectual and popular ideas very similar to social media platforms of today. Newspapers also included images and pictures, which give readers insight into the literal image of the working woman geologist from the Victorian period until the 1980s (Brenned and Hardt 1999).

"Glamourizing" the female scientist was not unique to the geosciences or the oil industry, but the synergy between the two fit war ends, nationalism, and the ends of some female scientists. In another field, such as physics, the newspaper became a vehicle for the promotion of women in science as a triumph against Cold War Russian forces (Kitch 2001). The *Painesville Telegraph*, published in Painesville, Ohio, published *Personality Portraits*, which profiled the life of Dr. Mabel Wilson, who was a professional chemist and researcher. Like many of the female geologists profiled in this book, especially those during the middle part of the twentieth century, Dr. Wilson was described as a high-achieving scientist, as well as a physically attractive woman:

Part of the important team of research chemists in Research and Development Laboratory of the Diamond Alkali Co is Dr. Mabel Wilson of 63 Chestnut St. In charge of the instrumental laboratory, she holds a responsible position and is well-known in her field as a capable scientist (Painesville Telegraph 1950).

The article then described her physical appearance, "When the tall brunette walked into the Diamond Research Building, December 1938, she was the first woman chemist employed by the Diamond Alkali." (Painesville Telegraph 1950). Ultimately, she was accepted because she transcended the skepticism of her male colleagues, according to the article, because of her strong will:

The reception from the masculine chemists was strictly formal and distant but after she proved her outstanding ability, she was readily accepted as a scientist and not merely as a woman (Painesville Telegraph 1950).

In the same manner as the female geologists profiled during the war, her hobbies, as well as her characteristics as a mother, are also profiled in the write-up of her work, "Striving to further knowledge in chemistry, Mrs. Wilson, the mother of two youngsters picked up her books and returned to Michigan State College where a dream of the past final came true." (Painesville Telegraph 1950). She was into knitting as well, "Away from the laboratory and in her home, Mrs. Wilson enjoys knitting and has made very known color combination of argyle socks for her son, Jack, 19. Television doesn't interfere with her knitting at all, she claims. She watches a program knits along as nicely as you please. Another favorite past-time of the lady chemist is gardening." (Painesville Telegraph 1950). The article ended with an emphasis on her real and lasting contributions to the world as a chemist, "Dr. Wilson might not have the income a medical doctor as she would have liked to have been but she and her fellow research workers are making contributions daily to help humanity and make life easier, healthier, and safer" (Fig. 1.1).[1]

Though this profile is similar to many of the geologists that will be discussed in this book, it does not strive for the heroic or pin-up glamour that newspaper profiles will emphasize in the 1930s until the women's liberation movement in the 1970s. The profiles of female scientists, especially those of female geologists, are at the intersection of women entering the scientific field professionally while having to navigate the standards of beauty and cultural expectations of womanhood (Cogan 1989).

An old argument, but a good and important one, is that woman in twentieth-century society were forced to master the technical and professional while being perfectly put together and perfectly made up. Most anyone would admit that those standards are still at play today. Female geologists during the early to mid-twentieth century were forced to be both the perfect scientist and the perfect homemaker. Figure 1.2 is a cartoon reminder for career women who need to maintain high standards of beauty and the household while still working and succeeding at a career. As my advisor Ann Johnson always emphasized, scientists both worked and lived in society and culture (Johnson 2009). This study shows how female scientists are affected by the culture they live in but also, in some cases, use those cultural values to fight for their place in petroleum geology (Dear 1995).

[1] This was transcribed as best as possible, as the print was faded.

Fig. 1.1 Wilson working in her laboratory (Painesville Telegraph 1950)

This study is not comprehensive, unlike Robbie Gries or other scholars, because I wanted to give the reader a thematic discussion of the role of women in both the culture of science and the overall culture they lived in (Gries 2017). Because of the proliferation of the Internet, social media, and more electronic-based portrayals of women geologists, this book could span thousands of pages. This book seeks to tell the first half of that story: the origins of the glamour narrative and its practice through the wars and into the end of the twentieth century. I hope that this book continues to open up more avenues for other scholars to pursue female scientists in social media into the present day. More books will likely be needed by other scholars to historicize the transitions of the new images of the working female petroleum geologist and the complex story of those individuals who identify as women. I also wanted this study to be useful for undergraduate history and science classes, as well as engaging for scientists, while being a manageable length. The book likely leaves out stakeholders and is not a comprehensive study, nor is it truly possible to write histories that are so comprehensive that they tell the whole story. Book-length histories are an attempt to tell an *aspect* of the story, and I would encourage curious readers to examine many of the other studies named in this book or write their own.

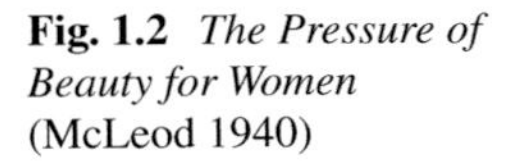
Fig. 1.2 *The Pressure of Beauty for Women* (McLeod 1940)

There simply cannot be enough of these books, as there are a multitude of stories out there. Other historians avoid scientific topics at their own peril and stick their heads in the sand. Historians need to write for historians but also for STEM audiences, as STEM audiences value history.

1.2 Newspapers and the Idea of "Glamour and Geology"

Many newspaper articles and narratives from the 1940s to the 1970s promoted the exciting image of the female geologist working in the field, often with a pickaxe, microscope, or rugged pair of dungarees. Women experienced many periods of transition. These transitions were captured in the pages of newspapers, periodicals, and scientific literature as women became an increasingly important part of the petroleum industry and geology in the United States. Portrayals of women in newspapers and especially trade periodicals made them larger-than-life figures and almost transcendent, but slowly women began to add accounts of sexual harassment, discrimination, and exclusion from science and industry. They had a type of superhero image portrayed to readers of newspapers and trade presses in the oil industry, and those heroes eventually added their own narratives, though the typecasting must have been exhausting, as women found themselves subject to expectations of the oil industry and larger society—to be beautifully dressed and simultaneously possess technical expertise in the field or geological laboratory.

The idea of what I will call the "oil woman," was a product of many different groups and stakeholders, including female geologists, the petroleum industry, the United States Armed Forces, and journalists. The concept of the oil woman remains as the literal title of a trade publication that features women working in the petroleum industry and is currently being published today. Major newspapers promoted the ideas of the glamourous, intelligent, and feminine geologist at work in the oil industry, and aspects of those ideas remain in the culture of science and the oil industry today.

Newspapers promoted women like June King, who they described as a "pin-up" and discussed her confidence, with her quotations reflecting her attitude about being a woman in a profession dominated by men: "A woman can do everything in geology that a man can do[.]" (*The American Weekly* 1945). King's likeness included in the article was in the pin-up style of photography of the time; see Fig. 1.3 (West 2020). She was portrayed as a "pin-up," and this portrayal reflected larger ideas about the roles and places of women in society, as female geologists and women who worked in the petroleum industry were affected by cultural, social, technological, and political changes occurring in the United States (Hegarty 2008).

Historically, I will argue that the concept of the oil woman had a type of duality that enshrined itself in the American mind, or imagination, during World War II. The woman working in oil prior to World War II was a woman moving into geological work because of wartime necessity but also drawing on skills that were deemed appropriately female, as geology was an educational subject that woman "of accomplishment" studied prior to the twentieth century, in addition to languages, mathematics, history, and other sciences. Geology was something that was popular and deemed feminine.

Kristine Larsen wrote in her study of women in the nineteenth century that women "...contribute[d] to the scientific endeavor[s] during this [nineteenth] century, in meaningful, if not less visible ways. Women were about to carve out well-defined niches for themselves, especially in the biological rather than physical sciences." (Larsen 2017). Larsen pointed out that there were many women who participated in botany, to the extent that it became a "feminized science." Female botanists received a lot of pushback when men tried to professionalize the field, as women were major contributors to the field as writers and collectors. Most of the middle-class gardens of pleasure and educational opportunities were open and cultivated by women. In fact, many science books in the nineteenth century, such as those written by Jane Marcet, were aimed at both women and children (Bahar 2001; Pandora 2009). Like botany, geology was accessible to women, according to Larsen, because it drew on the drawing skills women cultivated in their education and botanical pursuits. Field trips made for proper activity for women, as they served as a companion for a father, brother, or husband. Changes in fashion allowed women more opportunities and ease for participating in fieldwork, such as the advent of the "divided skirt or bloomer" (Larsen 2017: 18–19). Gathering rocks, even fossils, was seen as non-strenuous and acceptable amusement. Some middle-class women like

Fig. 1.3 Pin-up style images of June King from *The American Weekly*

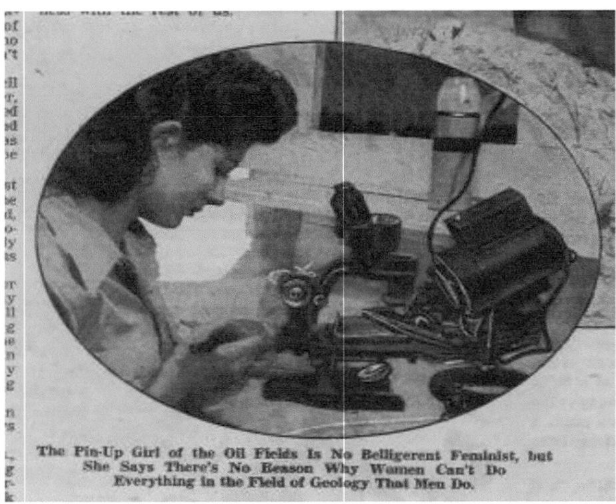

Mary Anning were able to make money from their collections and expertise (Larsen 2017). Other women made money through their geological writings. Though women could work in geology and produced geological knowledge during the nineteenth century, "…there were certain very influential doors that were firmly shut against them, namely the various learned societies" (Larsen 2017: 20).

One woman who exemplified the transition from the Victorian period of the accomplished woman to the working woman in field geology in World Wars I and

II was Augusta Hasslock Kemp. During her lifetime, Kemp, born in the Victorian period of the United States (1860–1918), found that a career in education, especially science education, was accessible to middle-class White women in places like Tennessee (Sheffield 2001). She received graduate training in geology at the University of Chicago. Kemp continued her interests in geology and served in various capacities in the petroleum industry, helping to discover oil in World War I and managing younger women who had received specialized training in petroleum geology during World War II.

A transition occurred in World War II in the United States when Kemp was put in charge of female geologists. Helen Bitgood, one of the members of the group, was profiled in newspapers as a symbol for changes in the oil industry, which the industry itself would continue to promote throughout its own publications. Bitgood was described as having a tough personality that could tolerate roughneck oil men and conduct sophisticated analysis of fossils through the microscope, but profiles also emphasized her domestic skills, marriage, and femininity. Think of geologists like June King and Helen Bitgood as types of "Rosie the Riveter" of the oil field.

Rosie the Riveter was a type of omnibus symbol for working White women during World War II.[2] After World War II, Rosie the Riveter became a feminist symbol for women's contribution, strength, and participation in historical events. Her symbolism carries additional meaning, representing not only the presence of women working and supporting the war effort but also highlighting the exclusion of Black women and other women's contributions from histories of World War II. Rosier the River is also a complex symbol, representing both female beauty and toughness, as well as efforts to perform male-dominated industrial work.

I also argue that women involved with the oil industry from 1940 to the 1970s continued the trends that started during World War II and were further *complicit in their own feminization*, as promotion by oil companies and newspapers helped them secure a firm place in the industry. Women later leveraged that image into roles as oil company owners, managers, or Chief Executive Officers. However, not all women chose the same path; some women or groups of women fought against the dominance of the oil industry by advocating for more rights and the freedom to choose their own dress or breaking away from the hyper-feminine image. Major social changes, such as the second women's rights movement and attempts to pass the Equal Rights Amendment, motivated women in petroleum geology to challenge the static image that had been portrayed from World War II until the 1970s (Berry 1986).

Other scholars, such as Ruth Milkman in *Gender at Work: The Dynamics of Job Segregation by Sex During World War II*, argued that neither women, nor labor unions, nor even professionals really put up much objection to the new "women's place," where women performed household tasks and jobs that required physical strength. Milkman contextualized the lack of resistance:

[2] See the excellent discussions of not only the history of Rosie the Riveter but also how to teach the history of women's labor during World War II in the *OAH Magazine of History* for Summer–Fall 1988.

> The wartime idiom of sex segregation combined such prewar themes as women's dexterity and lack of physical strength with an emphasis on the value of women's multivaried experience doing housework and an unrelenting glamourization of their new work roles. That "women's place" in wartime industry was defined so quickly and effectively owed much to the power of this sextyping. Although the initiative came from management, neither unions nor rank-and-file workers—of either gender—offered much resistance to the *general* principle of differentiation of jobs into "female" and "male" categories. Nor was the ideology of "women's place" in the war effort ever frontally challenged. There was a great deal of conflict, however, over the location and the boundaries between the female and male labor markets in wartime industry, and over wage differentials between these newly constituted markets, and this is the subject… (Milkman 1987; 64)

The oil industry was a special space for women, as there were female-owned oil companies in the twentieth century (Ponton 2019).

Throughout newspapers of the twentieth century, there was an excitement about women and technology (Stanley 1995). Women seemed to be at the forefront of science and technology in the oil fields. Newspapers often pictured women with instruments or new scientific interventions to promote the new and cutting-edge ideas of the oil industry. In 1947, for instance, as shown in Fig. 1.4, women were pictured with X-ray technology used to look for oil.

As much as women were praised in newspapers and by oil industrialists for their fieldwork, they received additional praise for their laboratory and advanced technological expertise, especially during World War II and the postwar era. Newspaper articles praised the technological accomplishments of women and linked them with the strategic success of the United States' war efforts. A 1941 article, for example, praised women performing research in laboratories as "doing their bit for Uncle Sam." The female reporter writing a series of articles about women working in defense research highlighted the importance of science and the war effort: "Democracy's life and death race for supremacy in the air is concerned with far more than factories humming around the block. No plane is better than the brains plowed into it in laboratories and drafting room. Women are at work in both." (Robb 1941). Dr. Hertha R. Freche was profiled for her work crafting better aluminum panels to improve war planes for the United States. Her research was deemed crucial, but, like many profiles of female geologists in the petroleum industry, she was praised for having, as well as doing, it all: "She [Freche] is literally one of those women whose work is never done. When she comes home from the laboratory she always has a piece of sewing, knitting, tatting or crocheting to pick up." (Robb 1941).

Other women working in wartime research, such as Dr. Katharine Burr Blodgett, were described as competent scientists, as well as having their physical features noted for the reader: "A slight, attractive woman, whose experiments have been liked in importance to Mme. Curie's, Miss Blodgett is the discoverer of 'invisible glass.'" (Robb 1941). She was also described as "doll-size" and down-to-earth, preferring to be called "Katie." While her scientific work was praised and her accomplishments included patents for glass technology, she was also praised for her work at home and her personality: "Dr. Blodgett holds a patent on a method of conditioning surfaces so that they are not wet by oil. This woman, whose hobby is gardening and whose bridge game is a Blitzkrieg, sews and embroiders beautifully, and her

X-raying Oil Fields

A new advance in petroleum research is this X-ray machine, which takes pictures of how oil, gas, and water flow thru rocks thousands of feet underground. The X-ray focusses on samples of oil well rock, into which fluids found below ground are injected. The resulting X-ray pictures reveal in miniature how oil behaves in various earth formations. An engineer in Gulf Research Laboratories, which pioneered the development, is shown adjusting a cylindrical rock sample, while a secretary prepares to record X-ray findings. The machine is expected to lead to greater percentage of oil recovery.

Fig. 1.4 Women using X-rays in petroleum exploration

darning is a thing of such geometric perfection that it ought to be framed instead of worn!" She was also credited with inspiring other women to work for General Electric and opening the workplace.

Write-ups of technological and laboratory mastery during World War II and into the postwar era often highlighted the hobbies, homelife, or feminine qualities of female scientists, such as those working in petroleum geology and other closely related fields. There was a pressure to have it all, accomplish major feats in the laboratory, and then run a household superbly. Car companies were aware of this working woman and focused ads targeted at these working women, who were also pressured to be equally dedicated to the home. The ad from *Life* magazine appears in Fig. 1.5 below.

Women had new opportunities in the oil industry and in petroleum geology, but they also found their domestic dynamics changing as well.

All of these changes represent the "oil woman" that I will discuss throughout the book. Visual sources will be examined in this book, in addition to traditional textual sources. Scholars like Ann B. Shteir and Bernard V. Lightman have argued that studying images related to science reveals not only important aspects of print culture but also larger points about the culture related to science and technology, such as education, gender, or the aesthetics of science (Shteir and Lightman 2006).

Fig. 1.5 Advertisement from *Life* promoting cars to women working in oil technology (*Life* 1950)

1.3 Understanding the Patterns and Meanings of Women at Work in Petroleum Geology

It is not a surprise that women felt discriminated against in the field of earth sciences during the late nineteenth and most of the twentieth century (Alic 1986). This is because many scientists still feel that there is significant inequality between women and men in science. There was also a continuation of racial discrimination in the sciences, as well as discrimination against those from lower socioeconomic levels and those identifying as lesbian, gay, bisexual, trans, asexual, or questioning. This book will focus on the discrimination faced by female geologists during the late nineteenth and early twentieth centuries. I hope this book will take its place among existing studies and those yet to be written about identity formation in science and the history of discrimination in scientific practice. Unfortunately, there are not enough histories of trans women in the fields of the history of science and geology, and I found that those sources were difficult to gather. I hope that similar studies focusing on trans scientists will be published in the future, as oral histories and other sources could be gathered or revealed through rigorous archival work.[3]

Scholars like Elizabeth M. Bass studied women and their contributions to the oil and gas industry in her dissertation *"That These Few Girls Stand Together": Finding Women and Their Communities in the Oil and Gas Industry* and found that women made up a small amount of oil workers: "Little has been written about women in connection with the oil industry." (Bass 2020). However, the women who were part of this industry made significant contributions. Previous studies, however, have had agendas, presenting discussions of women in the oil industry as celebrations without much context or "commentary" and focusing on empowering women to pursue a career with the momentum of historical legacies. Bass pointed out that:

> The purpose of these works is not only to celebrate the achievement of the women featured, but also to encourage young women and girls to consider a future career in the industry, following in the footsteps of the path-breaking women featured. But also to encourage young women and girls to consider a future career in the industry, following in the footsteps of the path-breaking women featured.[11] (Bass 2020, 4)

Bass differentiates her research by stating that her dissertation is not designed to "…serve as a laudatory recruitment tool for female involvement in the oil and gas industry, this dissertation examines two questions." (Bass 2020, 5). The first question is how women working in the oil industry affected their own experiences and how they created communities and networks in a predominantly male-centered industry. She focuses narratively on the "peripheral work" of nonspecialist workers, such as office workers and those in the service industries of oil "boom towns." One of the most important points of analysis that Bass provides comes at the end of her dissertation, where she argues that women developed their own networks similarly to men in the oil industry. In discussing the collaboration of women from technical

[3] There are several studies beginning to be published on trans persons and science history. For more information, see Black (2022).

and administrative roles, she notes: "As an added benefit, they found camaraderie through friendships and the work of various geological organizations—their own 'good ol' girl' network as a counter to the industry 'good ol' boy' network—that made the difficulties of forging a career path in a male-dominated field more enjoyable." (Bass 2020, 161–162).

Bass's own work expands on that of Robbie Gries. Robbie Rice Gries wrote the cyclopedic book *Anomalies—Pioneering Women in Petroleum Geology: 1917–2017*. *Anomalies* is as compendium of women who contributed to petroleum geology (Gries 2017). The book is extensively researched and is one of the largest contributions to the field, especially in terms of size. Gries aimed to tell the story of women in petroleum geology, as she was a petroleum geologist herself. The book started as an attempt to memorialize the contributions of women to the Association of American State Geologists, and she shifted her focus to commemorate petroleum geology's centenary, continuing her diversity advocacy as a member of the American Association of Petroleum Geologists. Interestingly, she had access to the membership cards of the American Association of Petroleum Geologists and believed that many women active in the gas and oil industry were members of the Association: "Another great asset to these membership cards was that they allowed me to illustrate numerically and graphically the history of women in the oil business. Though not every female geologist in the oil business was a member of the AAPG, most were." (Gries 2017). While Gries had a lot of biographical information from the cards, much historical information was unfortunately missing and not preserved in archives, leading to a frustrating series of dead ends, destroyed diaries, or simply no historical records: "After many disappointing dead end pursuits, I had to admit they were lost forever, it broke my heart." (Gries 2017). Same.

Gries needs to be recognized for her contributions and kindness to other scholars, as she was eager to share information, such as allowing researchers like me access to her files, appearing in interviews in Bass's dissertation, and sharing or mailing copies of valuable historical records that helped this book, such as Esther Applin's unpublished autobiography or the "Rock Puppies" piece from Dorothy Aylesbury McCoy. She did a tremendous amount of genealogical and biographical work as well, tracking down family members and distant relatives of some of the geologists who appear in her book. She was extremely helpful to me, and I appreciate it.

After examining the collection of AAPG cards that contained women's names, Gries then supplied information in a graph (Gries 2017, 6). Gries emphasized that throughout the era from World War I, the advent of micropaleontology, World War II, the postwar era, and the era of affirmative action, women in geology ultimately rose to larger amounts of participation. Her goal was to excite aspiring female geologists and to recover lost history (Gries 2017, 6). I feel that Gries really had a lot of passion for the project and wrote the book to understand her own place as woman in science and geology, which made for an interesting read. However, the study needed more cultural and historical context for what was going on, as Gries focused a lot on the intellectual contributions of female geologists as the field of petroleum geology was coming into its own maturity and profitability. The cultural turn in the history of science has been a sharp and intellectually profitable one, as many scholars have pointed out that the culture in which science is created matters quite a bit (Dear 1995; Terral 2006). Gries captures the struggles of what it meant to be a woman in a field that was hostile to women, but historians could add more to the narrative about what it means to be a scientific woman in the larger context of being a woman in greater society.

The pressure of being a pioneer is useful in showing the struggle of the larger number of women profiled in her book, but it can be off-putting and disengaging to students interested in science who feel they must be exceptional in all cases and do not experience the freedom of being an everyday scientist. This book moves away from the pioneering image of women in geology, as that has been adequately profiled, and instead looks at the everyday woman in science. It examines the construction of historical narratives that stress the everyday uniqueness of female geologists. Sometimes, female geologists and the oil companies themselves worked to construct an equally beneficial narrative so that they could both succeed. At other times, female geologists went along with the narrative, while at other times, they pushed back against it.

After reading many of the trade journals such as *The Petroleum Engineer* and *Oil Week*, I have the impression that, although the oil companies and sometimes the women themselves embraced the trailblazer and pioneer narrative, the concept of "anomaly" does not quite fit their appearance in the oil and gas industry. This is especially true given that at the same time the oil and gas industry was actively promoting female superheroes of oil during the war, women started to rebel against that narrative later in the century. The concept of anomalies that Gries uses as the narrative hook for her book was defined in the secondary title page as *Anomaly—A Departure from the Expected Normal: A Geologic Feature Which is Different from the General Surrounding and Is Often of Potential Economic Value*. I think a more apt description is **transitional figures.**

I would argue that women working in geology, especially petroleum jobs, and those in administration, camouflaged their work in the wider aims of industrialism or nationalism. Much of the work, or ideas about image, were adaptable and sometimes used to fit the best professional outcome. Some women deconstructed or unmasked that narrative in their desire for more autonomy to participate in larger political projects later in the century. One former title of this book was *From Mrs or Miss, or Ms. or Dr: Women and Petroleum Geology*, because women participating in geology had their identity focused on being the woman of accomplishment, and then opportunities and roles opened up to become highly technical and well-trained participants in geology. This transition is apparent from the life of Augusta Kemp to that of Ellen Bitgood, for instance. The stories of women are disguised in many newspapers, but the meaning of these articles not only contributes to further biographical collections of women but also speaks to the role the oil industry and newspaper writers were constructing for them, and in some cases, how those women were imagining their own roles in petroleum geology.

My own work also builds on the work of Gries and Bass by arguing that much of the identity formation of some of these female geologists, especially those from the World War II generation, benefited from the public relations work of oil companies and newspapers, which shows a shrewd marketing approach by the oil industry. Female geologists, like female autoworkers and industrial works in 1940s America, did not fight against the *feminization* of their profession and embraced many of the archetypes constructed in the 1940s (McEuen 2010). There was resistance of this impulse in the 1960s–1980s. The concept of the oil woman remains in the consciousness of the oil industry today and continues to impact marketing in magazines such as *Oil Woman*. Magazines like *Oil Woman* continue to push the message of the exceptional female oil worker blazing a path through a mostly male-dominated profession.

One of the points of this book, however, is that the contributions of women in geology should be of use to women studying science and aspiring to scientific careers. I think there is a middle ground between uncritical celebrations and historical analysis without understanding the value and stakeholders of the story. The value of the story is well reflected in the article "Exploring Science Identity Development of Women in Physics and Physical Sciences in Higher Education," written by Ebru Eren. In the abstract, the article sums up the importance of historical and social scientific research in defining the identity of women working in science, as well as the challenges women experience in science. Eren summarized that:

> Understanding women's science identity development help brings a general view about developing a more welcoming and flexible science culture for individuals who think they do not fit well or who are left outside of certain prevailing norms in the scientific climate. It also can allow seeking a way of challenging and changing the predominant culture and the prevailing masculine norms in doing science (Eren 2021)

Sharing narratives and histories of women in science challenges the assumptions and ideas about who does science and creates spaces of invitation for formally marginalized persons to join the scientific enterprise and community. Reading about the contributions of female geologists likely continues to expand the narrative of who participates in earth science.

Bass pointed out that as of 2020, only 16% of oil workers were women. However, many trade press pieces throughout this book historically argued that although women were exceptional in the oil industry, they were excellent. The identity of the female petroleum geologist, as well as office workers, needs to be examined to see how these women operated in the oil industry and how the oil industry portrayed their roles. We can historicize the nature of identity for women and see how internal and external forces shaped it.

1.4 Women and the History of Identity

Women's lives have been marginalized in history writing for many years. However, after the second women's rights movement in the 1970s, historians began to include the history of women in their analysis of historical events. The historian Natalie Zemon Davis wrote a history of three women who were on the margins of history

(Davis 1997). These three women were not well known in their time and are not well studied by historians to this day. The three women lived during the Western Middle Ages, and though they were not at the center of society, they made their own space and contributions. Davis, speaking of the women she chronicles, wrote that, "In their own way, each woman appreciated or embraced a marginal place, reconstituting it as a locally defined center." (Davis 1997, 210). These women, on the margins, made their own way in the Middle European world: "In each case, the individual freed herself somewhat from the constrictions of European hierarchies by sidestepping them." (Davis 1997, 210). Though not every woman written about in this book sought to overthrow the largely patriarchal nature of science or subvert the systems of power in place, each created her own world within a world and was reflexive about her place in the practice of geology in its various forms.

Davis, though writing about women who were Jewish and Christian, wrote that, "Each life stands as an example, with its own virtues, initiatives, and faults, and seventeenth-century European motifs run through them all…." (Davis 1997, 212). Each woman discussed in this book was a complicated individual who was affected by, and who affected, the society and culture in which she was working. It is the job of the historians to try, I would argue, to pull the lives of women together to say something holistic and explanatory about the triumphs, struggles, and challenges of being a woman in the laboratories, fields, classrooms, and worksites that involved geological knowledge and how those experiences were influenced by the person's gender.

Many historians have written about women in science since the early 1980s and have used history as a lens to explore current inequality in science today. The historian Sally Gregory Kohlstedt in her piece *Women in the History of Science: An Ambiguous Place*, written in 1995, stated that "Issues of equality and fairness are fundamental for host historians who explore the participation of women in science. The social and political circumstance of the last two decades helped spur the investigation of women in history even as other intellectual trends pushed an increasing number of scholars to investigate the history of science and technology, particularly in the modern period."[4] (Kohlstedt 1995, 39).

Other scholars have tried to understand why women are not equally represented in science, and often the answer is that there is a false narrative suggesting that the history of science did not include women. Scholars are not united about how to remedy problems in science regarding representation, with some feminist writers, though a small number, arguing that science should be completely overhauled to create a proper place for women (Keller 1987). There are many problems that need the full attention of various stakeholders, such as whether scientific education should be co-educational, the support of exclusive female academic institutions, and new findings about the psychology of learning (Sugimoto and Lariviere 2023). The nature of sex and gender could also be the subject of a book in itself (Schiebinger

[4] The typeface of this quote was changed from all caps to regular case.

2004). Finally, questions about retaining women in science are important, as most of the resources go into recruiting rather than retaining women (Golden et al. 2011).

I am of the opinion that history and history teaching can stop the leak of STEM majors and integrate science and the humanities further together in a working mission. In my previous book, *Early Nineteenth Century Chemistry and the Analysis of Urinary Stones*, I shared that one of my main interests in entering history is to improve science. Improvement certainly could include data and knowledge creation but also enhancing the workplace, climate of discussion, and experiences of knowledge producers. History can also be a tool for scientists to use in addressing problems in science. I hope that this book is just one in a long line of helpful collaborations between science and history. A profound hope I have for this book is that it changes someone's perception of women in science, particularly fostering an interest in the relationship among industry, knowledge, and the creation of the professional self.

Identity formation and the creation of the self is a complex process that often continues throughout one's entire lifetime. The creation of the self is too deep a concept for this book, or perhaps even several books. However, sociologists, psychologists, historians, and other scholars have found that many factors contribute to the construction of identity, such as one's culture, economic status, race, and gender. One important piece of scholarship on the history of identity comes from Louise A. Tilly, who recounts her experiences in academia (Tilly 1989). She stated that she attended a conference presentation where a scholar discussed the writings of women during the French Revolution, and a member of the audience questioned what the knowledge of women being present in the French Revolution meant in the grand scheme of things.

Tilly noted that this question emphasized the importance of connecting the history of women to larger questions in the field of history, not simply adding more information to existing fields. Adding gender into the history of women opened up a new form of analysis in history, and though women's history itself had been accepted into mainstream academic history by the 1980s, the concept of adding the analysis of gender showed how the experiences of women reveal underlying power structures and the importance of society in the construction and deconstruction of women's experiences. Tilly writes that "The introduction and spread of the concept of gender as a socially constructed category in historical writing has been an effective counter to biological determinism; it has strengthened comparison and the study or variation and process and, in its use in deconstruction, called our attention to power relationship." (Tilly 1989).

So, what does that mean? The answer is that advertising, music, films, literature, and other forms of material (manufactured items) and cultural expressions are both a mirror of what it means to be a woman in society and what it "should" be or what is *expected* of women in society. The history of women, with its start in "the rediscovery of biography and personal testimony as important history sources," gains additional context and meaning through analysis, which provides insights into the collective history data about women in fields such as petroleum geology (Tilly 1989). Both aspects, according to Tilly, are extremely important, and I would agree. Tilly concludes that writing about women in the French Revolution reveals the deeper struggles of different groups concerning power during the Revolution. In analyzing power, constructions of gender were important as "Gender was a metaphor for other unequal relationships as the struggle to consolidate power played itself out." (Tilly 1989).

The analogy with the French Revolution and women is important to understanding why it was crucial to study the role of women in the petroleum industry in this book. Women were negotiating routes of power and legitimacy throughout the lifetimes of figures like Augusta Kemp, a geologist working in the late nineteenth and early twentieth centuries, and June King during World War II. Working women geologists faced a male-dominated power structure in their industry, and to fight for their ability to produce knowledge, contribute, and make a living, they either aligned with some of the prevailing narratives or were complicit in the feminization of their profession to gain more access and power. As social movements changed in the 1970s, some women resisted the narratives promoted by the oil companies, while others maintained them to preserve their advantages. The concept of the "oil woman" and the exceptional biography in industrial narratives continues to this day in publications like *Oil Woman*; many of the stories appearing in industry trade journals retain the style of those written in newspapers in the 1940s.

1.5 Glamour, Geology, and History: Summary of the Book

This book explores how female geologists, often working in or involved with the petroleum industry, conceived of their identities as intellectuals, working professionals, geologists, and women. The book consists of 11 chapters that explore facets of identity creation. Each chapter delves into the world of female geologists as well as the world they created.

Chapter 2 explores the life of Augusta Thekla Hasslock Kemp, a Tennessee-born geologist whose life spanned the late Victorian period until World War II. Her life represents a crucial contextual point, as Kemp's life marked the beginning of transitions for both women in science and women in geology. Much like many of the characters in Larsen's book, *The Women Who Popularized Geology in the 19th Century*, Kemp was a woman of accomplishment, educated in geology and other subjects as part of a proper education for women. She pursued further graduate education at the University of Chicago. During her graduate studies, she faced sexism and difficulty conducting fieldwork as a woman but excelled nonetheless. Kemp continued her work in geology, moving to Texas and teaching high school, where she inspired future generations of students, both male and female, to appreciate earth science. Kemp is included in this book because she also contributed to the US war effort during World Wars I and II. She was involved in female geology groups searching for oil to support the US war machines. Her life is important as it represents the transition from high school classrooms and polished female education to working as a field geologist. Kemp likely mentored, or at the very least interacted with, many of the field geologists who entered the petroleum geology field during

World War II and the postwar period. Kemp was also incredibly creative and used literature and fiction to express her experiences as a geologist, woman, mother, and person living in late-nineteenth and early- to mid-twentieth-century culture.

Chapter 3 is a close study of the career outlook for women in geology. Many of the sources examined in this chapter were produced by the US government bureaucracy, especially the US Department of Labor, Women's Bureau, which created literature and pamphlets explaining the job outlook for women aspiring to enter the sciences, including geology. Colleges, universities, and professional groups encouraged women to build on the wartime momentum from World Wars I and II. Some women took advantage of wartime courses designed to quickly prepare them to work in the oil industry or produce maps. Women's sections in major professional geology groups emerged, writing about and describing the experiences of women in geology and geography. They also discussed the problems and difficulties women faced while performing fieldwork. Other pamphlets and literature produced by female geologists for female geologists promoted the excitement and versatility of a career in geology. It was advertised in several promotional pieces of literature that women were not limited to fieldwork, as they could also teach or work in laboratories.

The US government pushed and promoted women to be excited about the sciences, as several pamphlets included images that glamourized women working in the field of geology. Some of these images were taken from newspapers that also promoted the work women were performing in wartime geology. Stories of women doing adventurous and important work were meant to inspire younger women to pursue careers in science and geology. Though the number of women studying and working in geology was comparatively small compared to men, several educators (both men and women) wrote about the challenges of training women in geology and how to recruit and train more women for the field. Opportunities for women in geology, advertised by the US Department of Labor, Women's Bureau, included writing for outlets like magazines and children's books, in addition to industrial reports. It is likely that some women wrote and promoted other female geologists in magazines and newspapers, as many of the women profiled in this book come from newspaper articles written by women.

Chapter 4 is an examination of female geologists and the pin-up culture present during the World War II period of American history. Geologists like June King became larger-than-life examples of women defying work stereotypes. She was portrayed as exceptional in terms of technical expertise and beauty but also as a normal woman who could have hobbies and interests. Female geologists working in petroleum and oil refining became celebrities, often referred to as looking as beautiful as some of the most popular celebrities of the day. The best summary of this cultural trend was "Mixing Glamour with Oil." Many newspapers and periodicals ran stories of female pin-ups, complete with pictures, that advertised the adventurous and high-achieving female geologist of the World War II era. Other women, like Ellen Bitgood, appeared in newspaper write-ups discussing the "unusual" jobs that women took up during the war and the exceptional women working in geology and oil. Some of these newspaper stories did discuss the sexism and difficulties these women experienced both in the field and in the office, but these concerns were often minimized in the adventure and glamour narratives that appeared to readers.

The scientific and workplace achievements of these female geologists were often portrayed alongside their accomplishments as traditionally defined women, balancing children, housework, and marriages while working in geology. The working field geologist was both an excellent scientist and a "lady." In many ways, working female geologists were a type of wartime celebrity, and that celebrity would extend into the postwar period as well. Newspapers, as well as higher education in the United States, promoted and provided opportunities for women to enter into geology, science, and wartime petroleum production. Profiles stressed that women "weren't sissies," and oil companies like Standard Oil sponsored radio shows that often hosted female scientists, including geologist, and promoted their work as part of a long line of female contributions to the history of science. Publications in the oil industry promoted women working in petroleum and geology and featured many aspects of their working and personal lives. Though White women benefited from the war and postwar boom of women working in petroleum geology, African American women did not benefit in the same way. Some wartime publications promoted the contributions of women working in wartime industry, such as *Negro Women War Workers*, but there were very few portrayals of working African American women in geology. However, African Americans did participate in the petroleum field during the war. African American women working in geology appeared in a special racialized section in the *Employment of Women in the Early Postwar Period*. Black women did not enter geology in the same numbers as White women because many historically Black colleges and universities (HBCUs) did not have many faculty or courses focusing on geology. This issue was addressed shortly after the war in educational studies.

Chapter 5 is a continuation of the study of the imagery associated with females in petroleum geology. The chapter specifically focuses on the excitement and integration of the microscope and micropaleontology. Scholars like Robbie Gries have emphasized the increase in the involvement of women in petroleum geology due to major figures like Esther Apelin and the development of micropaleontology. This chapter explores the importance of micropaleontology in the rise of women in geology but focuses on the adventure narrative of women with microscopes and exploring ancient fossils with new tools like the microscope. Women with microscopes were a romanticized newspaper narrative in much of the postwar period. Female geologists finding uranium were portrayed in postwar newspapers as heroic and patriotic. Some of these women who were associated with this new geology became advocates for women in science and criticized the status quo treatment of women in academia.

Chapter 6 shifts the narrative of the book to women working in the oil industry and both participating in and constructing narratives about women in science. Women continued to be portrayed as pin-ups and heroic figures in science in publications like *Life* magazine in the postwar period. Oil companies picked up their aggressive marketing and recruiting of women for not only scientific positions but positions in other areas of the company, such as childcare, photography, translation, scientific writing, and art. The oil industry also worked with women to promote oil, science, and geology to community groups for women. These groups included the Desk and Derrick clubs that could be found all around the United States and Canada. This chapter examines the roles of women involved with the clerical side of the geology and petroleum industry. The Desk and Derrick Club was highly influential, as it appealed to women both inside and outside of the oil industry and gave out information about petroleum geology and other uses of petroleum-based products. Organizations like Desk and Derrick provided scholarships and other types of support for aspiring female geologists. The organization was involved in promoting beauty standards and further spurred the participation of women in diet culture. Newspapers of the time also profiled and reported on the experiences of women working in oil and geology. In the pages of newspapers, women also participated in spreading awareness of oil corporations, corporate-based propaganda, and narratives that emphasized the importance of oil in history. Columns, produced by female employees at oil companies featured clever characters, such as "Petroleum Peggy." Industrial publications like the *Oil & Gas Journal* had special feature sections entitled "Women at Work," where women across the oil industry, in many different types of jobs, were profiled to illustrate the presence of women in the industry.

Chapter 7 is a continuation of the study of women working in the oil industry, especially in the post-World War II period. This chapter dives deeper into office culture, journalism, and the jobs that women were doing that were not exclusively field-based geology. This chapter begins to explore the concept of the "oil woman" and its prevalence from World War II until today, as the concept is still used in trade publications. The "oil woman" was a concept that describes all the imagery, excitement, and newspaper articles that depicted the various roles women played in the oil industry. Both the industry and the women themselves seemed to embrace the image. The image was prominently featured in magazines for women, such as *Glamour*. June King, a working petroleum geologist, makes an additional appearance in the magazine to promote her role in the industry. The article and others like it were very much written with the "male gaze" in mind. The oil and gas industry seemed to willingly make room for women working in many different capacities, but sexism was still present, as women were valued for their technical accomplishments, while their physical beauty was often highlighted as an equally important part of any magazine or newspaper profile. Women profiled often had their beauty associated with their success in geology or the oil business in general.

Though some women went along with the promotion of ultrafemininity in the oil industry to gain access to technical jobs in oil, the industry sought active input from women working in the field and in the office. One example is the Women's Discussion Group at Mobil. The group was later renamed by a winning entry from a female employee as "Mobilogue" but was changed by the legal department as "Distaffforum." The group started out much like an internal Desk and Derrick Club, where featured speakers spoke about various topics related to the oil industry, national security, and other subjects of interest. The group was principally run by the female employees of the company. The purpose of the group was also for women to meet each other in the company and form a community. In retrospect, it was the formation of groups like Distaffforum that likely led to the organization of women in the company and allowed them to renegotiate their rights in the office. One example is the organization of women in an effort to be able to wear pants in the office. The ability, or freedom, to wear pants in the office was a proxy issue for women's rights and the extension of more autonomy and freedom for women at work. Newspapers and other trade presses covered the issue as a women's rights issue.

Prior to arguments over workplace attire, oil companies tried to convey expectations for housewives of employees. An internal publication of Mobil ran an article entitled "Are You a Handicap to Your Husband," which rightly infuriated many women associated with the company. Women wrote in to voice their concerns, anger, and frustrations. The editors of the magazine were forced to respond, unfortunately, with sarcasm. During the 1960s and 1970s, women associated with geology and the petroleum industry started to voice their criticisms about their place within the industry and argued for more equality and autonomy. Profiles of female geologists, though continually infused with causal sexism, recorded the women's thoughts on rights and cultural issues involving women.

Chapter 8 focuses on women within petroleum geology reacting to and expressing their ideas around the dominant cultural events of the 1960s through the 1980s, particularly the women's liberation movement. Women involved with geology also expressed their ideas, in newspapers, about their own bodies, sexuality, and issues associated with pornography. Profiles of women working in geology still had a pin-up style but included their thoughts and feelings about political issues of the day, such as the Equal Rights Amendment. Women trained in geology expressed their ideas in the political sphere, as they ran for public office. As the oil industry allowed women to work in more field-related jobs and interact with male employees, there was some cultural pushback, which was reflected in newspapers for a national audience. Female geologists were not silent on the issues expressed in advice columns like Ann Landers. They discussed cultural issues of the day, such as changing roles at home between men and women. Some female geologists even discouraged women from entering the field because of their experiences with exclusion and discrimination. This chapter examines the shift from women adhering to the oil industry narrative to working women in geology who expressed their experiences in petroleum, including negative ones. What started as small allusions to sexism and discrimination in some newspapers during the war became more pronounced during the 1960s onward.

Chapter 9 discusses the continually changing role of women in geology as the women's rights movement began to unfold. Women themselves advocated for more autonomy, independence, and rights in the office, including dressing for work and revising beauty standards. However, the oil industry and journalists responded to racial and cultural changes in US society by constructing celebratory and more inclusive narratives of women working in geology. African American geologists began to be included in the narrative describing women working in petroleum geology. Profiles of African American geologists, like Edith Williams, appeared in historically Black-focused magazines such as *Ebony*, where she was described as a "Lady Pioneer in the Oil Fields." These more diverse profiles can be seen as part of the new glamourization of geology. Women working in the oil industry also had to contend with economic issues, such as the massive economic crises in the oil industry during the 1980s, where several women lost their jobs. Some women did not hesitate to comment on their experiences with job loss in popular magazines.

Women also were entering management and commented on how women should properly manage employees in the oil industry. Ideas for women, by women, appeared in oil trade publications. However, one of the largest changes was that profiles of women working in geology and petroleum geology began to feature their struggles with discrimination and challenges working in the industry as women. These profiles contrasted with the celebratory type of profiles of June King. These women did not discuss their roles as homemakers but addressed new cultural ideas, such as more equitable relationships between husband and wife. More testimonials and journalism captured the overt sexism in the field that women had to experience, such as on drilling platforms. As negative stories proliferated in newspapers, magazines, and the general news cycle, the oil industry continued to run new and innovative profiles of women working in oil that countered this negative press.

Chapter 10 serves as a conclusion of sorts, but not an ending. The experiences of women working in petroleum geology continued to appear in newspapers and flourished in media. Women's toys, like *Barbie*, embraced the career path of geologist. *Barbie*'s appearance harkened back to the profiles of the postwar period that romanticized the fieldwork of a working geologist. Female geologists, who had experience in the petroleum industry, entered the space program and appeared on broadcast television. The idea of the oil woman returns in the appropriately named *Oil Woman* magazine, published for the interest of women working in the oil industry. Newer profiles of women working in petroleum geology, post-COVID, are more complex, reporting more on the whole woman profiled and her ideas and experiences up front in the article. Glamour continues in geology but is new, updated, and in the context of the interests of women.

References

(1945) Mixing glamor and oil. The American Weekly, September 30

(1950) Key member of important team of chemists in diamond lab is capable lady scientist. The Painesville Telegraphy. 28 October

(1950) My work trips to the oil field are so much easier with b-w overdrive! Life. 23 January p 50

Alic M (1986) Hypatia's heritage: a history of women in science from antiquity through the nineteenth century. Beacon Press, Boston

Arredondo et al (2022) Women and the challenge of STEM professions: thriving in a chilly climate. Springer Nature, New York

Bahar S (2001) Jane Marcet and the limits to public science. Br J Hist Sci 34:29–49

Bass EMB (2020) "That these few girls stand together:" finding women and their communities in the oil and gas industry. Oklahoma State University, Dissertation

Berry MF (1986) Why the era failed: politics, women's rights, and the amending process of the constitution. Indian University Press, Bloomington

Black R (2022) Transgender women trailblazers are changing the nature of science. 14 March: https://nhmu.utah.edu/articles/2023/05/transgender-women-trailblazers-are-changing-nature-science
Brennen B, Hardt H (eds) (1999) Picturing the past: media, history, and photography. University of Illinois Press, Urbana
Cobble DS (2015) The sex of class: women transforming american labor. Ithaca: Cornell University Press.
Cogan FB (1989) All-American girl: the ideal of real womanhood in mid-nineteenth-century America. University of Georgia Press, Athens
Davis NZ (1997) Women on the margins: three seventeenth-century lives. Harvard University Press, Cambridge
Dear P (1995) Cultural history of science: an overview with reflections. Sci Technol Hum Values 20(2):150–170
Driggers EA (2023) Early nineteenth century chemistry and the analysis of urinary stones. Springer, New York
Dyhouse C (2010) Glamour: women, history, feminism. Zed Books, London
Ebony MGO (2020) Interrogating structural racism in stem higher education. Edu Res 49. https://doi.org/10.3102/0013189X20972718
Eren E (2021) Exploring science identity development of women in physics and physical sciences in higher education: a case study from Ireland. Sci Ed 30:1131–1158
Foster AE (2011) Integrating women into the astronaut corps: politics and logistics at NASA, 1972–2004. Johns Hopkins Press, Baltimore
Golden M, Mason MA, Frasch K (2011) Keeping women in the science pipeline. Ann Am Acad Pol Socia Sci 638:141162
Gorham D (1982) The Victorian girl and the feminine ideal. Routledge, New York
Gries RR (2017) Anomalies: pioneering women in petroleum geology, 1914–2017. Jewel Publishing LLC, Lakewood
Hegarty M (2008) Victory Girls, Khaki-Wackies, and Patriotutes: the regulation of female sexuality during world war ii. New York University Press, New York
Honey M (1984) Creating Rosie the Riveter: class, gender, and propaganda during World War II. University of Massachusetts Press, Amherst
Johnson A (2009) Hitting the brakes: engineering design and the production of knowledge. Duke University Press, Durham
Keller EF (1987) Women scientists and feminists critics of science. Daedalus 116:77–91
Kitch C (2001) The girl on the magazine cover: the origins of visual stereotypes in american mass media. University of North Carolina Press, Chapel Hill
Kohlstedt SG (1995) Women in the history of science: an ambiguous place. Orisis 10:39–58
Larsen K (2017) The women who popularized geology in the 19th century. Springer, New York
McEuen MA (2010) Making war, making women: femininity and duty on the American home front, 1941–1945. University of Georgia Press, Athens
McLeon ET (1940) Beauty after forty. Pittsburgh Post-Gazette. 25 April
Meyer LD (1996) Sexuality and power in the women's army corps during World War II. Columbia University Press, New York
Meyerowitz J (ed) (1994) Note June clever: women and gender in postwar America, 1945–1960. Temple University Press, Philadelphia
Milkman R (1987) Gender at work: the dynamics of job segregation by sex during World War II. University of Illinois Press, Champaign
Morris VR (2021) Combating racism in the geosciences: reflections from a black professor. AGU Adv 2(1). https://doi.org/10.1029/2020AV000358
Pandora K (2009) The children's republic of science in the antebellum literature of Samuel Griswold Goodrich and Jacob abbot. Osiris 24:75–98
Pizulo C (2011) Bachelors and bunnies: the sexual politics of playboy. University of Chicago Press, Chicago

Ponton R (2019) Breaking the gas ceiling: women in the offshore oil & gas industry. Modern History Press, Ann Arbor
Robb I (1941) In research laboratories, shops, women contribute their share. St. Petersburg Times, May 11
Schiebinger L (2004) Nature's body: gender and the making of modern science. Rutgers University Press, New Brunswick
Sheffield SL-M (2001) Revealing new worlds: three Victorian women naturalists. Routledge, New York
Shteir AB, Lightman B (eds) (2006) Figuring it out: science, gender, and visual culture. Dartmouth College Press, Hanover
Stanley A (1995) Mothers and daughters of invention: notes for a revised history of technology. Rutgers University Press, New Brunswick
Stewart-Williams S et al (2021) Men, women and stem: why the differences and what should be done? Eur J Person 35. https://doi.org/10.1177/089020702096232
Stone PK, Sanders LS (2021) Bodies and lives in Victorian England: science, sexuality, and the affliction of being female. Routledge, New York
Sugimoto CR, Lariviére V (2023) Equity for women in science: dismantling systematic barriers to advancement. Hard University Press, Cambridge
Terrall M (2006) Biography as cultural history of science. Isis 97(2):306–313
Tilly LA (1989) Gender, women's history, and social history. Soc Sci Hist 13:439–462
Wallace A (2005) Newspapers and the making of modern America. Bloomsbury, New York
West M (2020) The birth of the pin-up: an American social phenomenon, 1940–1946. Dissertation, University of Iowa
Williams CL (2021) Gaslighted: how the oil and gas industry shortages women scientists. University of California Press, Oakland

Chapter 2
Augusta Thekla Hasslock Kemp: Woman of "Accomplishment" and the Pre-glamour Geologist

2.1 Changing Identities

Augusta Thekla Hasslock Kemp lived a life of changing identities. Born toward the end of the Victorian era (1820–1914), she saw America transition into the industrialized and progressive early twentieth century, and she passed way during the start of the sexual revolution in the 1960s (Driggers 2018, Sklaroff 2009, and Rupp 1997). It also must be noted that she passed prior to the 1963 eclipse in Texas (Kemp 1965). Ideas regarding gender and science were in flux during Kemp's lifetime. However, the notion of female scientists challenged the status quo of the role of women in the American society (Schiebinger 2004; Jardins 2010). Kemp was also working in geology without the imagery of glamour and in the pre-wars, pre-glamour period.

Kemp was not a clear radical or progressive reformer. She held many identities throughout her lifetime and alternative between them. An example can be found in a description of her work done by her sister, Clara Whorley Hasslock. Hasslock collected and published her sister's writings and oversaw the donation of her papers in Texas and Tennessee. In the introduction to the book *Pegasus Limping*, Hasslock writes the history of her sister's life noting that: "Most of her non-professional writings were done while she was teaching and keeping house." (viii) Hasslock further commented on the gendered identity that her sister moved through by publishing Kemp's thoughts on her own writing (Kemp 1965).

Kemp thought of herself as a geologist, teacher, wife, and writer. This is best exemplified in her poem "Why." She muses that:

> Why am I a good teacher? Because I study rocks in my spare time.
> Why am I a fairly good housekeeper? Because I teach.
> Why am I a good cook? Because my husband is such a fine "critic teacher."
> Why does my husband love me? Because I pay him so little mind.
> Why does he enjoy his home? Because he gets waited on, hand and foot.
> Why do I like words? Because I read Scott and Thackeray, Dickens and Stevenson.
> Why do old ladies like me? Because they enjoy my chatter.

E. A. Driggers, *Glamour and Geology*,
https://doi.org/10.1007/978-3-031-64525-9_2

Why do little boys like me? Because they bring me worms.
Why do I write this? Heaven only knows. (Kemp 1965, pg. 77)

In this chapter, I explore the complex dynamics of Augusta Thekla Hasslock Kemp identity. This chapter will focus on four facts of that identity: geologist/scientist, teaching, writing, and her relationship with her community, family, and artistic self. Kemp is a transition point between the woman of accomplishment in geology and to those working in the professional oil field. Her life trajectory embodies both the continuation of the feminization of geology and petroleum engineering and shows how women like Kemp mentored and affected the next generation of women working in the oil fields, especially during the World Wars. She is also an active scientist who not only published creative work, but many different scientific papers throughout her life (Kemp 1934, 1959, 1962 and Miller and Kemp 1947). The struggles that Kemp goes through during those transitions also reflect the larger social changes in society for women.

2.2 Learning and Practicing Geology in the Pre-glamour World

Augusta Thekla Hasslock Kemp grew up in Nashville, Tennessee. She was born on July 21, 1882. The late 1880s were the tail end of the Victorian Era. The Victorian Era could be summarized by separate spheres in labor in regard to scientific and intellectual work. Women could be educated in ornamental sciences, such as mathematics, languages, histories, and geology in order to make them proper companions. Teaching and education were seen as appropriate careers for women, as well as working as "computers (Light 1999; Rossiter 1980)." Pursuing science in carefully constructed spaces would have been as reasonable aspiration for a woman of the class and background of Kemp.

An obituary summarizing Kemp's life appeared in the *Baylor County Banner* published in Seymour Texas on August 1, 1963. The obituary named Kemp in relation to her husband, calling her "Mrs. J. F. Kemp Passes." The obituary described Kemp's educational credentials. Kemp graduated Peabody College (now part of Vanderbilt University as its College of Education) in Nashville in 1901, receiving a teaching degree. Her sister's biography described it as a "Licentiate of Instruction Diploma." In 1902, she received a Bachelor of Science degree from the nearby University of Nashville (now also part of Vanderbilt University) (Kemp 1965; Conkin 2002; Baylor County Banner 1963).

After teaching in the Tennessee Public Schools, she began teaching in science classes, which included physics, chemistry, and geology in a high school in Abilene, Texas. After spending time in the education system in Texas, Kemp traveled to the University of Chicago for her Master of Science degree at the University of Chicago. The timeline for her studies has become confused by some scholars and historians (Vance 2001; Kemp 1853–1974, 1882–1963, 1965). Kemp's transcript was included in her personal papers at Texas Tech's Southwestern Collection.

The transcript and papers show that she began her studies at the University of Chicago as of June 17, 1905 (Kemp 1853–1974, 1882–1963, 1965). Some historians only include the 1910 date of her thesis acceptance, but she was at work as a graduate student in the Summer Quarter of 1905, and she continued to take exams and have her field work recognized by the University of Chicago until July of 1936 (Hasslock 1910). While at Chicago, many of her professors wrote of her accomplishments in the program. She continued her interests in geology and furthered her education by taking courses in "…summers in the Universities of Wyoming, Colorado, Montana, Missouri, Chicoga [sic Chicago], Iowa, and Texas." (Kemp 1853–1974, 1882–1963, 1965).

Her Master of Science degree was formally awarded on June 14, 1910 (Kemp 1853–1974, 1882–1963, 1965; Hasslock 1910). She took many of her classes in the summer during the first three years of study, without any classes in 1906. The classes that she took in summer quarters from 1905 to 1908 included physics, zoology, and field geology, which included labs (you can see this on her transcripts as well). Her grades ranged from A to the lowest being a C, when she took "The Pedagogy of Physics" as a "visitor." Kemp started going full time to graduate school by the Autumn Quarter of 1909, where classes that she took included the "Influence of Geology on American History." She took classes ranging from mineralogy to "Commercial geology, Advanced Course," and passed her exam in geography on June 6, 1910. There were several examples of Kemp taking correspondence courses and exams in her papers, and she continued to pursue advanced geologic education throughout the course of her life (Kemp 1853–1974, 1882–1963, 1965).

Kemp supported herself and her family by teaching, mostly in high schools in Tennessee and Texas. She also taught across the West in Arizona and Oklahoma as well. She married the principal, John Franklin Kemp, while she was teaching in Abilene, Texas, in 1914. There were two wedding announcements that were saved in Kemp's personal papers. One was very traditional that praised the appearance of the bride, "The bride was lovely in her wedding costume of embroidered net and shadow lace. The long tulle veil was caught in a Juliet cap with clusters of lilies of the valley. Her bouquet was a shower of lilies of the valley and bride roses." However, the second announcement described Kemp differently and commented more deeply on Kemp, her character, and her job situation (Kemp 1853–1974, 1882–1963, 1965).

The second announcement started with Kemp quitting her job: "Miss Augusta Hasslock has resigned as science teacher in the high school and will be married September 1." (Kemp 1853–1974, 1882–1963, 1965). Her husband was described as having a "fine salary," "brainy," and "honorable." The article alludes that their relationship began in college. Kemp is described as "Miss Hasslock, who classes herself as a 'plain' woman and even makes jobs about herself, is one of the finest characters that ever stood before a class in the high school. She is highly intellectual and yet it has not over balanced a rich store of good, hard, common sense." The article further describes her as "Her education, it is believed, was largely acquired through her own efforts, showing the strong character and determination that acut[sic] her life. She has very womanly trait strongly marked, and those who had come to know her best are her most ardent admirers." It also noted that all her friends approved of the relationship and they appeared to be happy. But Kemp did not quit her teaching job and continued teaching in Texas for many years (Kemp 1853–1974, 1882–1963, 1965).

John Franklin Kemp was prompted in 1920, and they both moved to Seymour, Texas, where they both taught (Kemp 1883–1943). Augusta Kemp later became the head of the department of science in the high school until she retired in 1943. One activity that constantly stands out in people's description of her life is her work in the "Explorer's Club" that was active at Seymour High School. Augusta Kemp mentored and provided extracurricular instruction to students regarding geology and other sciences. She took students on field trips. Kemp also had private students that she tutored in German, as well as other sciences (Kemp 1853–1974, 1882–1963, 1965).

Kemp received several honors over the course of her lifetime. She was inducted to the Texas Academy of Science as an honorary fellow. She was maintained membership in such scientific organizations as, "...Texas Geology Association, the American Association of Petroleum Geologists and Mineralogists, the Geological Society of America, and the American Association for the Advancement of Science." She donated her personal fossil collection to the University of Texas and other fossils and materials to the Texas Technological University. The obituary ends with an attempt to frame Kemp's character: "Although Mrs. Kemp possessed a brilliant mind and was well known in scientific circles over a wide area, she never lost her common touch with young and old, who continued to bring her specimen and to consult her on the ordinary insects, plans and rocks of Baylor County" (Kemp 1853–1974, 1882–1963, 1965). One of the most important parts of her legacy is her students, who the obituary notes, "...are to be found today in positions of responsibility and service." The testimony of Kemp's students will be a vista into her identity as a geologist and an educator (Kemp 1853–1974, 1882–1963, 1965).

2.3 The World of Augusta Kemp: Victorian Geology Transition for the Twentieth Century

Augusta Kemp's training and career in geology and the earth science was taking place in a shift of science and the education of women. This chapter's main theme is how a woman, born and becomes of age in the Victorian Era, transitions from a woman of accomplishment and professional opportunities mostly in education, to a working geologist who is the beginning of a wave of women who start to work as field geologists, such as those women profiled at the end of this chapter. Kemp's career trajectory takes places during the late Victorian Period. The Victorian period is named so because it took place over the lifetime of Queen Victoria Queen of the United Kingdom and Empress of India, who lived from 1837 to 1901. However, those trends and cultural values continued to stretch into science and culture, and some historians would argue that the Victorian age truly ended at the end of World War I in 1918. Historians involved in childhood education in the United

Kingdom define the Victorian era as lasting from the Napoleonic Wars until the start of 1914 (Goodman 2014).[1]

The Victorian Era, at the same time, has been increasingly difficult to define, as some scandals in research have emerged in the scholarly historical community, especially in regard to sexual ideas and beliefs (Goodman 2014). Some people in the United States hold a very narrow definition of the Victorian Era as a time of extreme social conservatism and manners, much like the seemingly oppressive and conservative nature of older historical conceptions of Puritanical America. However, neither of those analysis was true. The Victorian Era was a time of gendered and separate spheres, but there were erotic novels, pornographic pictures, explicit movies such as Thomas Edison's *The Kiss*, sexual devices circulating through the email, and access to birth control. The literature of the time, such as Bram Stocker's *Dracula* and Charles Darwin's *Origins of Species,* not only opened the world up to science but also, implicitly in the case of Darwin or explicitly with Stoker's work, explored human sexuality (Russett 1991; Ward 2014; Larson 2006).

Scientific literature was circulating and expanding during the Victorian Era (Second 2014). Verity Burke discussed the popularization of geology by female writers (Burke 2022, pg. 660). She summarized that during the Victorian Era that, "Women writers were also active in popularizing geology, authoring text ranging from posthumous biographies of prominent geologists to poetry" (Burke 2022, pg. 660). Burke historicized that women became popular authors for other women and children regarding books on scientific topics. The idea of the respectable and accomplished woman was still in vogue, and woman read works in geology and other sciences as a type of ornamenting, in the same way that they would practice music or art.

The analogy from the book *Pride and Prejudice*, though published in 1813, almost 100 years prior to Augusta Hasslock attending college, continued to remain in ideas about femininity and the proper role of women in society. Charles Bingley, an aristocrat, who along with the romantic interest in the novel of Mr. Darcy, defined the woman of accomplishment: "A woman must have a thorough knowledge of music, singing, dancing, and the modern language, to deserve the word; and besides all this, she must possess a certain something in her air and manner of walking, the tone of her voice, her address and expressions, or the word will be but half deserved" (Austen 1813/2010, pg. 73–74). Darcy added to the definition: "…and to all this she must yet add something more substantial, in the improvement of her mind by extensive reading" (Austen 1813/2010, pg. 73–74).

Augusta Hasslock was coming of age between the accomplished woman and the new woman in American and in the United Kingdom that was pursuing high education and scientific training (Burek and Higgs 2007). Burke added that scientific training for women was deemed suitable as long as it served as a functionary of their domestic duties or execution of female virtue, as the characters and structure of

[1] If a historian tries to say that the Victorian period ended prior to World War I, they should immediately be viewed as untrustworthy.

female composed scientific works "…featured activities and experiments performed in a domestic setting under the audience of a mother-figure, were considered respectable ways for writers such as Jane Marcet (1769–1858), Delvalle Lowry (1762–1824) and Maria Hack (1777–1844) to publish on geology" (Burke 2022, pg. 660). However, some female writers wrote for adults as well and received popular consumption and praise (Lightman 2007). Women like Augusta Hasslock could have access to these books, as well as publications aimed at women such as *Every Girls Magazine*, which contained scientific articles and writings (Burke 2022, pg. 660).

Hasslock also found herself in the middle of major changes in scientific education and higher education for women. Scholars such as Margaret W. Rossiter have pointed out the dilemma that many women found themselves in during the late nineteenth and early twentieth centuries. While higher education, as well as scientific education, was expanded in the 1880s to 1910, women did not have easy access to employment after completing their education. Rossiter summarizes the difficulties for women like Hasslock: "When the movement to give women a higher education had begun to take hold in the United States in the 1870s and the 1880s, little thought had been given to the eventual careers that such graduates might take up" (Rossiter 1980, pg. 381). The education, therefore, would led to the women being better wives and mothers: "Because of the prevailing notion of 'separate spheres' for the two sexes, most women were assumed to be seeking personal fulfillment and to be planning to become better wives and mother" (Rossiter 1980, pg. 381). But the expectations began to change as the number of women increased in colleges. Rossiter argued that not only were female scientists advocating for more participation of women in the scientific labor force but there were larger forces shaping increased calls for scientific training and job accessibility (Rossiter 1980, pg. 382). Unfortunately, many of the jobs in science that opened up to women were both low status and low pay. Women were also funneled into working with children or detailed oriented position. Many of the positions open to female science students were those in the female sphere, or what were called "pink collar jobs" later in economic and historical literature: social service, teaching, healthcare, and childcare (England and Boyer 2009).

Though Hasslock had a large amount of scientific training at the Peabody Institute in Nashville, her course of study was focused on becoming an educator, as she chose to attend this institution for a suitable career, along with male students. She chose to become an educator which had high status, especially in the communities that she served as a teacher. The job in education allowed her plenty of time for fieldwork and pursuing creative pursuits. It is unclear what her prospects would have been if she had chosen a non-conventional route going straight into industry or continuing for a doctorate. Her pursuit of a master's degree was successful. It is interesting that none of the female scientists whom she was in charge of during World War II pursued a career in education, given the time period (Terzian 2006).

Hasslock was deeply in between the women of accomplishment and those who chose a career in education and the women entering industry as geologists.

During 1898–1899, Kemp was attending high school in Nashville. High school education for anyone during the late nineteenth and early twentieth centuries was rare and equal to a higher education degree today. She took math, English, history, and history, in addition to music. From an early "Report of Written Examination," many of Kemp's grades by subject are preserved. During the first half of her final year in high school, as high school was only eleven grades at the time, she scored a 52 in math. The grading scale was that 90 was "excellent" and 80 was "very good" and 70 was deemed "good," but anything less than 50 "forfeits place in grade." However, Kemp pulled her math grade up to an 85.

She was studying for her L.I., or Licentiate of Instruction, at the Peabody College, now Vanderbilt University. This was a requirement to teach in Tennessee, but this degree has been discontinued. She stayed for her Bachelor of Science at the same institution. Interestingly, when she retired in 1943, her announcement in the paper named her as a BS and MS holder but did not mention of her earlier educational credentials (Kemp 1853–1974, 1882–1963, 1965). The degree of L. I. stands for "Licentiate of Instruction" as the degree was discontinued in many states during the World War I era (Beach and Rines 1912, Pulliam 1968). She received her L. I. degree in 1901 and got her bachelor's degree in 1902. Her Peabody transcripts were requested in 1940 and she took many science classes from plant histology to biology for teachers, mathematics, physiology, drawing, humanities (including many history classes and classical studies). She and her classmates are pictured below in Fig. 2.2. She often had a habit of taking candid and casual pictures, expressing her free spiritedness. The education included music, educational theory, physic, and domestic cooking. While taking a heavy load of science courses, she published creative writing pieces in *The Peabody Record*. Kemp posed for a picture in 1903 after graduation with her extended family (Corkin 2002).

The notion of womanly virtue and moral character is equally important in job applications for women of Kemp's time. Feminine virtue is equated with scientific ability. In June 1910, after completing her Master's Degree at the University of Chicago, one of her professors and examiners, Rollin D. Salisbury, wrote her a letter of recommendation to work at the Winthrop Normal and Industrial College (now Winthrop University) in Rock Hill, South Carolina, and described that he recommended her to teach geography and geology. He described the then named Hasslock (her surname prior to marriage) as "Miss Hasslock is an excellent student and efficient woman." His letter then continued to praise her professional qualifications: "I have no hesitation whatsoever in commending her to you, and in stating that, in my judgement, she can give efficient instruction in any subject which she is willing to under take." He ended the letter by continuing to describe her credentials as a woman: "Any institution which secures her will secure an able, efficient, right-spirited woman" (Kemp 1853–1974, 1882–1963, 1965).

2.4 Geologist and Educator

The geologist and physician W. Floyd Wright wrote into the *Baylor County Banner* to send his condolences and share his memory of Augusta Kemp. He wrote that in his time at Seymour High School he did not realize the high quality of his education at the time, it was only in retrospect that he realized the great opportunity he had in high school,

> When I was in Seymour High School, like many other students, I did not know that she was one of the best science teachers in the state. Not until I was in college did I appreciate what an outstanding scientist and teacher was Mrs. Kemp. She could have taught with qualification and ease in college and universities...As a person, a scientist, and a teacher, she was an inspiration. (Kemp 1853–1974, 1882–1963, 1965)

The write up that W. Floyd Wright, who would later receive his Doctorate of Medicine from the University of Colorado, was not the only student to highlight the skill of Kemp.

Glen McDaniel, who eventually rose to the executive ranks in companies such as Litton Industries and RCA, included a memoir of his time studying with Augusta Kemp. McDaniel was a student in her science classes at Seymour High School. McDaniel was a student around 1925–1928. He specifically focused on her time leading the "Explorer's Club." McDaniel recounts a story of Augusta Kemp discovering the Nautiloid that would be named for her:

> Between 1926–1928, Mrs. Kemp organized the Explorers Club, a fossil hunting group, of which I had the honor of being the President. Our first venture into the hill was on the [McDaniel adds the note T. A. (Coler)] property east of Seymour where we found a number of fragments of nautiloids. These were of the same general type of Permian Nautiloids as the species later discovered by and named for her, the Kightoceras Kompae. (Kemp 1853–1974, 1882–1963, 1965)

Augusta Kemp's greatest contribution to geological scholarship is her work on nautiloids and her work that focuses on the soil geology of West Texas.

In a memoir by Alfred S. Romer, an accomplished geologist, who was one of Kemp's professors at the University of Chicago, Romer commented on her zeal in collecting specimens. Interestingly, Romer describes her complex identity. He describes her as a teacher: "Her majority lifelong occupation was the teaching of high school science, notably in the Seymour High School from 1920 to her retirement in 1943" (Romer 1964). However, it was her pursuit of a graduate degree that was fueled by her love of the "Permian beds of her home region and their fossils—particularly cephalopods." It was her zeal to collect that was highlighted: "She collected invertebrate fossils with unflagging energy, and accumulated a valuable collection" (Romer 1964).

Romer wrote that her legacy was that she left a large collection of fossils to the Texas Bureau of Economic Geology (Romer 1964). Ultimately these collections of over five thousand fossils were distributed to the University of Texas at Austin and Texas Technological University. Kemp gave her collection of about five thousand invertebrates to the University of Texas, and after hear death, according to the University of Texas Alumni 1964 edition of *Alcalde*, as well as many out of print books and journals (University of Texas Alumni 1964).

In February 1957, Kemp published "Color Retention in *Stenopoceras, Euomphalus*, and *Naticopsis* from the Lower Permian of North Central Texas" (Kemp 1957a). Kemp was specifically in the Maybelle Limestones in the Southwestern part of Baylor County. In this paper, Kemp describes some of her fossils from Baylor County in Texas. She starts her paper by describing the color of the shell as indicative of their identity. She writes of the *Stenopoceras* that the single specimen she found have "...growth lines which make this specimen unique, for some of them are darker shades of light brownish red and some are lighter" (Kemp 1957a, pg. 974). She also found that it had splotches that appeared on its shell, much like the *Nautilus Pompilius* (Kemp 1957a, pg. 974). Next, she did comparative work by examining eight of the *Nautilus Pompilius*. She found that similar lines of growth had cream and "pale pink" colors. The article then transitions into discussing the E*uomphalus* genus. Kemp writes that she found these specimens in several sites around Texas: Talpa and Grape Creek, Elm Creek, and the Admiral foundation. Most of the specimens Kemp discusses come from Baylor County, except for grape Greek which is located in Thorckmorton County. She describes the largest amount of the specimens she examined from her collecting work in regard to their color. Vividly, Kemp describes the shells:

> The main mass of the shells is light colored, gray to tan. The darker pigment is diffused through the outer shell layer of the last who whorls, or two and one-half to three, in the larger specimens. The color is light brown to brownish black, being almost black along the carinae and the outer whorl face, especially on the last whorl. It is lighter and brown on the basal side. This pigment is completely soluble in dilute acid. (Kemp 1957a, pg. 975)

Kemp's careful analysis clearly and concisely conveys comparative facts and information to the reader. It must also be highlighted that her extensive collecting allows her a breath of specimens.

The final genus, the *Naticopsis*, is compared last. Kemp was able to examine only about ten percent of the "hundreds" of specimens that she collected. She notes that they do show the descriptive color of *Naticopsis*, but not "conspicuously" (Kemp 1957a, pg. 975). They went from gray to an extremely dark colored gray. Other specimens varied from darker to an almost "light tan." Sample plates from all three of the nautiloids are included with color plates and appear in Fig. 2.1.

Kemp ends the article by noting that she has other examples of naticopsids that are not mentioned in the article. She concludes that the distinctive line colors of the naticopsids are consistent in both young and old aged nautiloids.

2.5 Educational Experience and Training and Education at the University of Chicago

Augusta Kemp wrote a short story about her time at the University of Chicago. She studied under Dr. Rollin D. Salisbury, who was the head of the geology department while Kemp was studying there. She wrote of his bon mots (smart remarks). These were both: "...a livid mixture of lightening and red ink, stirred by a thunderbolt were both the joy and terror of his admiring students, respected with awe, and in

Fig. 2.1 *Naticopsis* plates. (Kemp 1957a, b)

after years, with love" (Kemp 1965, pg. 134–135). Kemp fondly remembered (after sufficient time had passed). She recounted such a bon mots from the first week of her graduate classes. The two had a repertoire:

> "How do you know a fossil sand dune?"
> "By the material of which it is composed," I answered.
> "Perfectly true, perfectly general, perfectly meaningless."
> But the shock did not paralyze me. On the country it stimulated me.
> "Sand!" I flung back.
> "Why didn't you say that at first?"
> Such was my accolade. They use the boys' slang term—I had been "crowned." (Kemp 1965, pg. 135)

Kemp noted that she had the proper skills to explain to Salisbury other things about geology. For instance, she was able to explain to him that the maps of how deep the ocean was were on page ten in his own book, not on the back on the book, where he assumed they would be located.

Being a woman in a mostly male program, Kemp had difficulties in finding female partner to share accommodations at field camps. She shared the story of such difficulties: "When I wanted to register for the field course the next term he sought, and found, a partner for me, a mature woman of ability who made a splendid field companion" (Kemp 1965, pg. 2). Kemp was also impressed that Salisbury remembered her by name when she returned to school after summer recess.

Kemp did well at the University of Chicago. During the 1909 academic year she had a "graduate scholarship, where she worked at the Geology Library. However, she was pressed to labor quite hard in the Geology Library, she remembered that "And I did work!" Although she said, "Most graduate scholarships were sincecures, even the undergraduate's were too, except for the routine of handing out an occasional book," but the geology faculty pressed the librarians. She remembered that: "The usual users of the books hunted up their own, but the Geology faculty conspired against the Devil to keep me out of mischief" (Kemp 1965, pg. 135).

As mentioned in the background and general introduction, Kemp was able to read German. Salisbury made her organize and file German sources. For instance, she helped another faculty member's course on Conservation, Harlan H. Barrows. She worked on keeping the slides for Wallace Walter Atwood. She also did research finding engineering articles for Emmons. Kemp also expressed her found remembrance that whenever T. C. Chamberlin desired for her to get a book, she lamented that "…everything stopped till the book was found. Not because he was an autocrat!" Kemp framed her own administration for him in gendered terms: "His [Chamberlin] patriarchal dignity, his old-fashioned stateliness, his courteous manner, his kindly, though somewhat absentminded smile, endeared him to all" (Kemp 1965, pg. 135). Kemp at the University of Chicago was not afraid to stand up to male authority, such as in the case of the book with Salisbury, but also was socialized to have an admiration for it.

Kemp also describes the interactions she had with some of the faculty members. She shared another story of Salisbury from a prospective of gender. In Walker, Salisbury was an often aloof faculty member but did "…occasionally departed from this custom" (Kemp 1965, pg 4; Boyer 2015). She wrote of a time that she encountered Salisbury in the Walker building, which housed both geology and geography, while working on examining a "specimen" in their mineral lab. The exceptional nature of a woman studying in the geology department was highlighted in Kemp's memoir:

> "Playing with rocks, Miss Hasslock?" he smiled at me.
>
> The amazed faces of the boys presented at the table told me how unusual that was. Another [sic] time one of the men and I were in a deserted laboratory; as a matter of fact we [sic] were studying together for our Master's examinations. Salisbury passed through, caught the apparently romantic set up, and grinned at us, a regular Cheshire cat grin.[2] (Kemp 1965)

Though Kemp does not bring attention to herself, these comments shed light on the social situation of attending an almost exclusively male program in a very patriarchal society. Kemp's memoir of her time at the University of Chicago continues to expound upon her memories about being a woman in the program as she ends the memoir of her time there (Boyer 2015; Kemp 1965). For instance, she highlights the difficulty of working after the library closed:

---

[2] Hasslock apparently corrected the manuscript to emphasize the grin imagery.

> The properties were strictly enforced; the library officially closed at ten P. M. If a girl desired to stay longer she got another girl to stay with her. This was not difficult, as most of the graduate women in the department lives in Beecher or Green residence halls, just a stone's throw from Walker. (Kemp 1965, pg. 5)

She further highlighted her difficulties working after hours. She notes the inequality of facilities access between men and women:

> Saturdays I usually worked on the fossils I had brought with me from Texas. Dr. Weller had given me a little corner in the Paleontology laboratory. The building was locked at one P. M., but someone managed to return before that time, and then had the job of letting the later comers in. Though men had entry to Walker on Sundays, it was "taboo" for the women.* (Kemp 1965)

Kemp included a note to further highlight what her use of taboo actually meant. The note at the end of the memoir makes clear that: "There was almost no girls or advanced students in the Geology Department and few in the Geography Department so that I often found myself thinking in that direction" (Kemp 1965). Kemp framed the inequalities between the small number of female student and large number of male students in both departments in very modest terms.

2.6 Women, Geology, and World War I: Analogous Experiences of Dorothy McCoy and Augusta H. Kemp

One woman who was a contemporary of Kemp was Dorothy Aylesbury McCoy. McCoy (often referred to as Mrs. Lewis McCoy) pursued an undergraduate degree in geology at Washington University (Gries 2017, pg. 11–12). She married another petroleum geologist, Lewis McCoy, and later worked for her husband in the petroleum industry. Historians like Robbie Gries have uncovered personal accounts of women like McCoy, who summarized their experiences during the World War I era (1914–1918), and family members of McCoy have shared these personal papers. The write-up of McCoy's experiences in the petroleum industry during World War I was named "Pebble Puppies" as that was the nickname given to women working in geology (Gries 2017, pg. 11–12).

The write-up of McCoy's experiences was an extraction that was taken from her personal diary by Everett Carpenter. Carpenter, who worked with Lewis McCoy, hired women to work in the petroleum industry around World War I. The "Saga of the Pebble Puppies" mentioned that most of the geologists of military service age were called to the War effort. McCoy was working in the Empire Gas and Fuel Company (which had changed its name in 1963 to the Cities Services), and McCoy was living in Oklahoma, and the company was understaffed. McCoy memorialized that "...practically all of their younger geologists were taken by the armed services" (McCoy 1963).

Dorothy McCoy framed the problems of World War I, which lasted from 1914 to 1918 in the United States, where the United States was involved in an international conflict between European allies. The United States fought alongside the Empires of the French, British, Japanese, and Russians, against the alliance of Italy, the Austro-Hungarian Empire, and the German Empire. Dorothy McCoy saw the problems of men serving in the war for the gas and petroleum industry, which was of significant strategic importance: “During the First World War, one of the industry’s big problems was the maintenance of an adequate working force in the face of the heavy military demands for man-power” (McCoy 1963; Stevenson 2013).

Dorothy McCoy was the first woman in an “experiment” the Empire Gas and Fuel Company ran to make up for wartime needs. McCoy was a working geologist, and covered needs in the office as well. Her service at the company was deemed successful, and more women were hired. These additional female geologists included Barbara Hendry and later Helen Souther. Souther and Hendry were trained at the University of Chicago. Another woman, Florence Travis, was also hired. These women were each given a portion of territory (McCoy, Souther, Hendry, and Travis) that stretched from Kansas to Oklahoma. They managed the oil fields in this large area of territory (McCoy 1963).

One of the most interesting parts of McCoy’s account is the field training that the four women received from the company. McCoy highlighted this additional training and break from traditional office work roles: “Aside from the regular office routine, the girls were required to undertake extra training in the form of several field training trips to give them first-hand knowledge of geology as applied to the petroleum industry; also, they were required to take instruction in [sic] practical pursuits such as map-making, beginning with the use of rod and transit in field problems” (McCoy 1963). The group also had a small community, as they attended professional meetings, such as the American Association of Petroleum meetings in Dallas of that year. They were accepted as “associate members” (McCoy 1963).

McCoy offers comments into how they were treated by male colleagues. She wrote that, “In consequence, since geologists were known as ‘rock hounds’, the girls were frequently referred to as ‘pebble puppies’ by their associates” (McCoy 1963). Gries noted that the different names “…indicated something about women not being equivalent to a ‘Rock Hound’” (Gries 2017, pg. 12). The history of the term is interesting as it persisted into the 1950s and 1960s, which it was used to encourage women to get into the field of geology. By 1921, in the *National Petroleum News* 1921 issue that covered the annual meeting of Petroleum Geologists in Denver, equated it with age,

> It was a highly successful meeting. There were geologists from everywhere, along with petroleum engineers and executives of a number of the more important producing oil companies. There were geologists from New York, form Los Angeles, from Canada and Old Mexico. There were geologists grey in years of service, and there were “Pebble-pups” just from their Alma Mater. There were company geologists, independent geologist, consulting geologists, and even insulting geologists. (Hazlett 1921)

The term would continue to change over time, but there is a point to be made about the inherent sexism and ageism in geology at this time. Kemp used the term “rock hound” in her poetry and writings later on. The women also received automobile maintenance and operation training. Because of the World War, McCoy emphasized that this was "instituted as a war measure" and it was somewhat of a novelty for the four women, as "...in those days many men, and very few women, knew nothing about driving an automobile!" (McCoy 1963).

The wartime work of the four women came to an end around 1919, when women started leaving the group to start families or pursue other employment. McCoy explained what happened to each member of the group. Travis left to take another job in 1919, then McCoy left to run the office of Alex W. McCoy, as he was working as an "independent consulting geologist" (McCoy 1963). Helen Souther left to marry another engineer, and Barbara Hendry left in 1921 to take up an opposition at Phillips Petroleum Company. Dorothy herself married Laws McCoy, who was not related to Alex W. McCoy in 1921. McCoy needed the entry by noting that in 1963 she was living in Oklahoma and many of the women had lost contact with each other. She signed the entry on August 1, 1963. (McCoy 1963).

Kemp's wartime experience was likely very similar to that of Dorothy McCoy. In geology publications from the World War I time period, Kemp is cited as contributing to the Bureau of Economic Geology. In the University of Texas's Alumni communication publication, *The Alcalde*, there is a memorial to her and her donation of 5000 shells (mollusks) from her time at the Bureau that she worked at in the 1920s. (University of Texas Alumni 1965) In 1918, in the *Geology of Runnels County*, Kemp is thanked for her access to her collection. In the acknowledgements, Kemp is thanked by her abbreviated name "A. H. Kemp" (Beede 1918). Kemp was cited for her work at the Bureau of Economic geology and her help with identifying well samples (Beede 1918). Unfortunately, the historical record does not fully capture her contributions at the Bureau of Economic Geology and her work in geology during World War I.

In regard to her ethnicity, Augusta Thekla Hasslock, came from German ancestry and this might have affected her experiences during the War. However, there is little documentation beyond her entries in *Pegasus Limping*. A section titled "German Writings" shows Hasslock's language and literary talents with the German language. There were several transitions of German writers such as Goethe. Hasslock also pursued additional training in German and tutored students privately in German. Kemp saw the pursuit of humanities as integral and just as important as scientific pursuits, as demonstrated in *Pegasus Limping* and the creative aspects of her work (Kemp 1965).

2.7 Being Creative: Framing Geological Work for Children and in Stories

In 1965, August Kemp's sister, Clara W. Hasslock, gathered many of the ephemera, stories, art, and student memoires regarding her sister. She published the book-length collection as *Pegasus Limping*. In her forward to the book, Clara Hasslock wrote that she had gathered all the "non-professional" writings of her sister, of which Augusta Hasslock had named "*My Idiot Children*" (Kemp 1965). Clara Hasslock introduced her sister as a geologist in her editoral entry. She was also described as a "teacher of the sciences," and a person who had other interests, like a creative side. She was someone who loved mysteries and literature and was someone who was very religious. The name of the book comes from a note in one of her writing notebooks that she thought of her creative writings as befitting the name "Pegasus Limping." Clara took that name and applied it to the book. After the publication of the book, it was promoted across local Baylor Texas radio and newspapers (Kemp 1965).

The book included an introduction, which is very interesting to understanding how Augusta Kemp's friends and family thought of her identify as a woman who was interested in science and creative pursuits. The biography promoted her educational accomplishments and teaching success, as she made money by winning competitive scholarship contests by examination and was a science student at the Peabody school, while simultaneously teaching in public school in Nashville, Tennessee.

In charting the narrative of her life, her public teaching career continued in Abilene, Texas. A couple of years after teaching high school science, she traveled to Chicago to pursue her masters degree, and four years after that her marriage to John Franklin Kemp was mentioned. And then from 1920 to 1943, she taught high school science. It was also mentioned that she was a private tutor in German and science until she died. And she was an avid rock collector, as visitors often brought her specimens to add to her collection. The introduction fondly recounted that, "Her former students often brought her rocks, as did the neighborhood children, for her to identify" (Kemp 1965, pg. vii). She continued to acquire geological knowledge, as the introduction pointed out, attending summer schools and field work at other universities. She continued her language training so that she could read geology articles in French and German. The introduction also pointed out that she contributed to the scientific literature in several journal articles and corresponded and sent reprints to other geologists around the world.

In *Pegasus Limping* she explored the role of nature and people and became very interested in American Indians. Hasslock did extensive traveling in the Western part of the United States and took a scrap book of pictures of her experiences meeting American Indian groups. She composed the short poem, "The Primitive in US," exploring the nature that there is something elemental in all human beings, "The primitive in us does live/We like the smoke of glowing oak./With jow we sniff to get a whiff/Or scorching ham in a camp-fire pan" (Kemp 1965, pg. 21). Kemp argued that all humans are tasked with searching for food and ends the poem on a bit of humor: "To catch a trout and pull it out./But when icy feet slick snakes do meet/We let out a yell and depart pell-mell" (Kemp 1965, pg. 21).

The creative work gives us interesting insights into how Kemp saw herself, the natural world, and her professional identity. In the entry "Poems About the Moon" after a description of the moon, she offers a type of personal connection: "If the moon came down to your gallery/And sat in the old arm chair/Would your big dog how, or run off the prowl,/And the puppies come up and stare?" Then she mentioned the moon in its relationship to science, "Drafted, a willing conscript into Science teaching." She then continued recording observations about the moon: "A modern monolith—a radio beacon tower." And finally, describing the moon's role with that of the earth: "The end of the rainbow withdrew from the earth" (Kemp 1965, pg. 23).

Hasslock wrote about her desires and experiences as a geologist. In "Geologists' Paradise," she spelled out what she considered a perfect situation for a geologist in the afterlife:

Heaven holds some promise
For talented musicians,
Since angels with their harps will sing
And play at all positions
But geologists will far the best,
So listen every rock-hound,
The very streets are gold they say,
So carry your hammer when heavenward bound,
Gold dust or agglomerate,
Gravel or conglomerate,
Sand or standstone, gold dust
Fresh or metamorphosed,
Gold that's smooth or glaciated,
Nuggets large, and much striated,
But if gold your steel may curl,
See the very gates are pearl!
But stay not for argonite, though
It gleams with heavenly light,
See the wonderous walls so bright
Chrysoprase, and jasper red.
This the Paradise for which you wept
Even when on Dana you slept. (Kemp 1965, pg. 27)

The poem describes her ideal practice of geology. The line "And play at all positions" might have been a longing for women to be able to practice all types of science at all types of positions. She also listed her preference in heaven to be a geologist, where she could carry a hammer to collect rocks. All types of rocks would be amiable for a geologist, or "rock-hound" to explore. She also discussed Dana, which might be the Irish goddess, which was also spelled Danu, which was associated with the earth mother or nature.[3]

Augusta also wrote about her life and larger global events that she lived through. She also made comments about her experiences as a woman. Entries include memoirs of World Wars I and II, as well as her experiences attending the World's Fair of 1893, and hearing about the Spirit of Saint Louis. She also commented on the areas where she was living, as she was highly interested in painting literature paintings of the geology. In the piece "Winter Solstice," Kemp describes the lands of Middle Tennessee. The part of Middle Tennessee she wrote about, likely about the Nashville area where she was born, grew up, and pursued her undergraduate education. She wrote of the end of the railroad tracks going east, where the hills became

[3] See MacKillop (2016).

progressively steeper, and the land was filled with lush vegetation like honeysuckle, which often grew on the occasional fences. The soils were different colors ranging from black to red, which came from the old coastal plain when Tennessee was beach front millions of years ago. There were thick clusters of trees that had refilled previously logged areas. She also described the Highland Rim area, covered in sand and oaks. Many of the names, such as the old Central Basin of Middle Tennessee, come from historical roots, as "The names descend from an early day in the history of Tennessee geology but they are just as appropriate today" (Kemp 1965, pg. 25).

Kemp also mentioned the hard and difficult labor that is associated with pursuing scientific questions. She wrote in "The Creative Spirit Based on a Laborious Compiling of Facts" of the slow and difficult research associated with science,

> Just as in the early stages of a summer storm, vapor condenses into minute droplets, each with a minute statist charge. As these droplets coalesce, their surface become relatively smaller and the density of their charges increases. Finally there comes the great illuminating flash of lightening. So in scientific work, the laborious computing of facts may require weeks, or months, or even years, still finally there comes the flash of scientific truth. (Kemp 1965, pg. 27)

Kemp likely experienced agrions chemical and geological research at the University of Chicago and in the field in Texas.

The book was also filled with memoirs to the losses she experienced during World War II. One entry "Pale Mottled Shells" was a story that Kemp used to explore her grief after the death of a friend in World War II. Kemp's sister, the editor of the book, wrote a footnote about the death of Benjamin R. Wirz, who died in action during World War II. The memorial sketch referred to the shells that Wirz collected, likely for Kemp:

> Gathered by a flyer on the distant beaches
> Of that far from peaceful ocean,
> For mother; sweetheart, sister, friends.
> The little brothers show the knives
> Strange shaped and queerly fashioned
> From the missing flyer's locker.
> The little sisters show the shells.
> And I, I think of the son, brother, friend,
> My friend, forever young, and alas, forever "missing."

It is likely that this is Second Lieutenant Benjamin Reinke Wirz, who was born in Seymore, Texas, where Kemp eventually settled. Wirz was part of the Red Raiders, a group of B-17 Flying Fortresses, also named the 22nd Bomb Group.[4] He served as a navigator in World War II and was Missing in Action over the Philippines. He is lamented in additional entries in the book. Kemp also includes a selection from the poem "In Flanders Fields" in 1940. It is likely that Kemp was experiencing a lot of loss, as 22,000 men and women were killed in World War II who were from Texas. This likely evoked memories for Kemp of the loss of lives in World War I.

---

[4] Please see https://www.findagrave.com/memorial/44687695/benjamin-reinke-wirz for brief biographical information.

Kemp was also creative with the camera, as she took many photographs during different stages of her life. Her archives in Lubbock, Texas as filled with photographs. She took many of the photographs and many of the preserved photos are candid, featuring family, friends, and romantic partner. Many of the photographs are self-portraits, and in context, Kemp was creating her own glamour. Two images that stuck out from my research in Lubbock include the images in Fig. 2.2.

The picture on the left is Kemp walking among corn in a moment of freedom and fun, while the picture on the right is Kemp, older, gathering either plant or mineral specimen in the field. The images that she included in her thesis also convey excitement and curiosity.[5] Like others who came of age in the Victorian era, there are several pictures of American Indians and their homes, likely taken on her field work and trips outside of her Texas home (Clemmons 1995; Wagner 2016; Moses 1996).

Fig. 2.2 Two portraits of Kemp (1853–1974). (Courtesy of Southwest Collection at Texas Tech)

[5] I encourage future researchers to examine the many photos included in the archives in the Southwestern Collection in Lubbock. Kemp has several photos, papers, and archival information at the Texas Tech Southwest Collection. A lot of the archival sources for this biographical chapter come from Texas Tech. Please see Guide to the Augusta Hasslock Kemp Papers, 1896–1964; Guide to the Augusta Hasslock Kemp Papers, 1865–1971 and undated; Guide to the Augusta Hasslock Kemp Papers, 1853–1974 and undated; Guide to the Augusta Kemp Papers, 1896–1965; and Guide to the Jun Franklin Kemp Papers, 1888–1943. Many of the photographs are lost and appear throughout the collections. Also thank you to Tom Vance and Jim Flis for the support and sharing of information about Kemp.

2.8 Creativity in Kemp's Geologic Writing: Exploring Her Master's Thesis

Augusta Thelka Hasslock completed her Master of Science in in the Department of Geology in 1910 at the University of Chicago. The "dissertation" for her Master of Science is titled "The Geology of the Area of Southwest of Abilene, Texas." (Hasslock 1910). Her dissertation shows an interest in the creative activity of photography. Thesis reads a thoroughly detailed deep description of the land, rock formations, fossil, and flora and fauna. The thesis is very similar to late Victorian naturalism with its emphasis on deep and detailed descriptions. Hasslock also included historical notes toward the end of her thesis, combining history with the creative actions of photography and detailed descriptions.

Hasslock also comments on the racial makeup of the people that inhabited the Western region of Texas in Abilene. She commented that the Mexicans who were involved in cotton cultivation and harvesting were called "Greasers," who were "from a very low social stratum in Mexico." She also noted that there were few African-American in the region (Hasslock 1910).

The thesis ends with her speculation of the "Future of the Region." She argued that Abilene would be mostly agrarian focused for the foreseeable future. She noted that the farming needed to be improbable, although irrigation would be out of the realm of practicality. She criticized the low-income farmers in the region with the criticism that: "The best development will not come till the number of tenant farmers is reduced, and the slipshod methods in bogue are changed" (Hasslock 1910, pg. 76).

She ends the thesis with an acknowledgement of the multidisciplinary, descriptive approach she took toward her scientific work. She summarized the region she studied in the thesis as, "This region is a very interesting one, geographically, whether viewed from the historical economic, or biological stand point, but the present discussion is merely a sketch" (Hasslock 1910, pg. 76).

The thesis was not done without assistance. She acknowledges that the field work for the thesis was completed while she was teaching in Abilene. She thanked the three other women who accompanied her while doing her field research, as a Miss Montgomery, and a Misses Cockrell of Abilene, who lived in Abilene. There was also a Miss Hester and Mrs. Hibbets from Merkel, who joined her. She described their company in her thanks that "...whose companionship made the field work possible, I am under deep obligations" (Hasslock 1910, pg. 11).

For historical context, around the same time that Hasslock was performing her field work, stories ran about a female geologist leading expedition into the field. One example occurred at the University of Chicago with the work of Zonia Barber. The article "Woman Geologist: Heads Expedition of Thirteen From Chicago University" (Baltimore Morning Herald 1902) described Baber as: "....considered one the best instructors in the department at the university, and being a fearless woman she received charge of one of the several groups now leaving for field work." Barber was accompanied by another female professor Ira D. Meyers of the education school. The group was studying the geology of Lake Superior (Baltimore Morning Herald 1902). In 1905, Harvard was advertising select field work courses for women, such as three weeks of field work that graduate female students could attend in glacial geology, held in New York (Boston Evening Transcript 1905).

Papers like "The Historical Problems of Travel for Women Undertaking Geological Fieldwork" by C. V. Burek and M. Kolbl-Ebert outline the many challenges and difficulties of women completing geologic field work, as these were significant factors hindering their work (Burek and Kolbl-Ebert 2007). Burek and Kolbl-Ebert summarized the tremendous challenges for women in the field: "From unsuitable clothes to lack of chaperones, from sexual harassment to lack of proper funding, throughout history women geologists have encountered difficulties traveling to their field, locations or working in the field, whether these locations were close by or abroad" (Burek and Kolbl-Ebert 2007). The authors emphasize the importance of field work with the maxim: "It is well-known saying within geology that the best geologist is the one who has seen the most rocks" (Burek and Kolbl-Ebert 2007). Previously, field work was considered an activity for men, as it was believed to be incredibly physically taxing and an aspect of heroic masculinity. The authors historicize that women began traveling into the field with their geologist husbands in the middle of the nineteenth century as chaperoned small groups of female geologists began to go into the field. Examples of such trips include the United Kingdom based Geological Association's activities. Then female students began to take field trips as well. More and more women were slowly increasing in fieldwork with colleges and geologists prospecting for oil. By the 1950s, women started doing field work for higher education and the oil industry (Burek and Kolbl-Ebert 2007; Burek and Higgs 2007).

Hasslock likely had to take her female friends and colleagues along with her to complete her field work because of the threat of male violence but also for proper social convention. Hasslock was also exceptional in doing field work. There are letters between her and other scholars that praise her field work.

In 1908, Hasslock received a reply from one of her professors at the University of Chicago. Samuel Wendell Williston (S. W. Williston on the letter) was replying to her letter describing some of her initial findings and ideas that would later make up her thesis (Martin 1994; Hasslock 1910). Williston was later thanked by Hasslock in the acknowledgements in her thesis, who praised her field work and her geological analysis. Williston wrote that "You have made some very interesting andimportant[sic] discoveries. You are unquestionably right in referring the upper beds to the Cretaceous. The Jurassic we know to be waiting throughout all of Texas, while the Trias has been reported from various parts of the west and northwest." He confirmed her findings that, "T[sic]e co lor [sic] of the rocks, their position all indicate with certainty, I think, that the deposits from which you obtained the prints are Triassic in age." He also praised the novelty and discoveries of her work and their important implications to paleontology and geology, "Theyare[sic], also, as cleary the prints of small, salamander-like amphbians—Microsauri or Branchiosauria or

Urodela. Now, the interesting fact is that we don't know of the existance of a single of of thesecreatures throught the world from all of the Mesozoic aga, save one single specimen from the Walden of Belgium and some from the uppermost (Laramie) Cretaceous..You discoveries I think will I fill out a very interest gap in ourknowledge" (Martin 1994; Hasslock 1910; Kemp 1853–1974).

He had received Hasslock's prints and says that he will analyze them but will wait to do so "…until we have gone over the horizon in early September. Where the[sic]se printshare so abundant there should be the impressions or remains of the skeletons, and a single skeleton of one of those little creatu[sic]s would be worth scientifically a month's search." He then ends the letter promising to join her in the field: "I shall be there in early September, and anticipate an interes[sic]sting and fruitful time in going over the beds[?] with you[.]" Hasslock was also thanked for sending along specimens. In his 1914 book, *Water Reptiles of the Past and Present*, Williston makes no mention of Hasslock and her fossil finds in Texas but does discuss dinosaurs and other fossil finds in Texas (Martin 1994; Hasslock 1910; Kemp 1853–1974).

Another professor she thanked in the acknowledgments of her thesis was R.D. Salisbury (Rollin D. Salisbury), who was previously mentioned in this chapter and in *Pegasus Limping*. She repeated a description of him as, " He was a near with a heart of Gold" (Kemp 1965, pg. 136). Hasslock used Salisbury's outlines of geology to teach her earth sciences classes in Seymore (Kemp 1853–1974). Salisbury was writing to Hasslock while she was living in Nashville. There are a lot of unclear points about when Hasslock was living in Texas or Tennessee. In his 1908 reply he noted that the letter was received and he was approving of her field work proposal. Salisbury replied that, "With reference to the field work which you were proposing in Texas, I have to say that perhaps you had carry it out successfully." But he cautioned that passing her thesis depended on the report, "Whether it can fount for credit, and whether the results could be accepted as a Master's thesis, would [sic] [d]epend upon the report" (Kemp 1853–1974). Salisbury included some advice about how to carry out her field work successfully, such as a careful examination of the beds and their rocks. And she would need to compare the fossils in the beds to others: "The proper completion of the work, for a Master's thesis at any rate, should involve a determination of rather full selections [sic] of fossils from the various beds." He recommended that she focuses on the "escarpment" which was a slope near the "northwest" and "central longitude" of the "Sweetwater sheet" (Kemp 1853–1974).

Salisbury fully agreed to mentor her field work and did not provide any caution or advice regard to her sex: "If you want to undertake this work, I am willing. I think it would be best register for course 20, namely the second field course" (Kemp 1853–1974). He also recommends another professor to advise her on her questions related to fossils. Salisbury ended the letter by mentioning that Samuel Wendell Williston will also be in Texas, and he recommends that they should connect or visit if it was convenient, which she does via the letter mentioned above.

2.9 Kemp and World War II: Forming Communities and the Transition to Glamour Geology

The headline from a March 2, 1947, edition of the *Wichita Daily Times* ran the provocative "Man's World in Oil Field Vanishes: Five Feminine Geologists Listed Here" (Wichita Daily Times 1947). The article continued its provocative nature stating that the work of the female geologists changed men's assumptions about women working in petroleum geology. The article opened that,

> Like the recovery methods of old- the once-boastful man's world of the oil field has vanished, and certain calloused souls who esteemed themselves masters of the surrounding topography have, on occasion, been compelled literally to eat sentiments not admissible in polite circles. (Wichita Daily Times 1947)

The article went on to discuss all the hard work that the women had undertaken during the war, as they were "On call at all times to check cores, read logs, watch cuttings, and predict the probably course of events up to and including the arrival of the latest hopeful into the Ellenberger lime…" The article also mentioned that the female geologist was equal to their male counterparts "…several of them have been official qualified to match their masculine counterparts in perversity over rutted trails and bogs to rigs where their expert opinions may be required" (Wichita Daily Times 1947). The article even went on to imply that the women were doing the job better than the men. The article went on to discuss the toughness of the geologists, as well as their talents,

> Defter than many a male at changing tires or extricating an automobile from hub-deep mud a mile from the nearest highway, these feminine geologists have long since established a reputation for the dependability and professional efficiency that few males, equipped merely with galoshes and a thermos bottle, have been able to challenge. (Wichita Daily Times 1947)

Kemp was referred to as the senior member of the five-member geology crew. The other members included Ellen P. Bitgood, Margaret Persons, Ethel M. Davis, and Beverly McMahon.

Ellen P. (Posey) Bitgood was from Petrolia, who had been working in the office since 1943. She was trained as a geology with both a bachelor and master's degree from the University of Oklahoma. She was described as having exceptional skills as a "consulting" geologist. She had previous work experiences with other oil companies.

Joining Kemp and Bitgood included "Three Latecomers"—Davis, McMahon, and Pearson. Ethel Davis was described as having an undergraduate degree in geology, as well as a master's degree after studying at the University of Rochester. She had previously taught undergraduate geology in the Pennsylvania-based Bryn Mawr College. She also had worked as a petroleum geologist in Texas. Beverly McMahon had worked at Shell prior to coming to the North Texas geologist corps in 1944. She had pursued geological education at the University of Colorado, achieving a bachelor's degree in the subject. McMahon had experience working in the oil industry with Shell in the labs. Finally, Margaret Pearson had worked for Standard Oil in Texas in 1945 and had education from the Ohio State University, after graduating with an undergraduate degree in 1942. She had worked in the United States Army Map Service prior to joining Kemp's group (Wichita Daily Times 1947).

Each woman was described, starting with August Hasslock Kemp. The article noted that she was now retired, was working in Seymor, and continuing geology research, including preparing a forthcoming paper that was going to be published in *The Journal of Paleontology*. After reviewing her educational background, she had previously worked for the Bureau of Economic Geology, which was located at the University of Texas in Austin, and had served as a "research fellow in geology" at the University of Iowa, in addition to many other memberships in professional organizations (Wichita Daily Times 1947).

In the pictures presented in the article, there is the beginnings of the associations of women working in geology and it being a glamourous and adventurous job. Unfortunately, their mentor is not pictures, and though she made her own glamour in her own photography, is not pictured in the article. Kemp or a family member noted her presence in the newspaper with a simple "X." The mark can be noted after the words "Senior Member" in Fig. 2.3.

Man's World in Oil Field Vanishes

Five Feminine Geologists Listed Here

ETHEL M. DAVIS

Senior Member

Three Latecomers

MARGARET PERSONS

ELLEN P. BITGOOD

Cooke Enj

Record O

Fig. 2.3 Kemp and women working in geology during World War II. (Wichita Daily Times 1947)

The next chapters will focus on the second generation, such as these women mentioned in the article, of women that Kemp likely mentored and examined their own identity formation and experiences in geology. August Kemp severed as a link between the Victorian period of women who learned geology as an accomplishment and the transition of women into the work force of geology, such as petroleum engineering and geology related to the War effort and post-War Period.

References

Austen J (1813/2010) In: Spacks PM (ed) Pride and prejudice: an annotated edition. Harvard University Press, Cambridge

Baltimore Morning Herald (1902) Woman Geologist: heads expedition of thirteen from Chicago university. 29 July

Baylor County Banner (1963) 1 August. Mrs. J. F. Kemp Passes.

Beach FC, Rines GE (1912) The Americana: a universal reference library. Vol 16. New York: Scientific American Compiling Department.

Beede JW (1918) The geology of runnels county 1(1). The University of Texas, Austin, p 6

Boston Evening Transcript (1905) School and college: field work in geology: program of courses established by twenty-eight colleges and university. 25 March

Boyer JW (2015) The University of Chicago: a history. University of Chicago Press, Chicago

Burek CV, Higgs B (eds) (2007) The role of women in the history of geology. The Geology Society, London

Burek CV, Kolbl-Ebert M (2007) The historical problems of travel for women undertaking geological fieldwork. Geol Soc 281:115–122. https://doi.org/10.1144/SP281.7

Burke V (2022) Geology. In: Morri E, Scholl L (eds) The Palgrave encyclopedia of Victorian women's writing. Springer, New York, pp 658–661

Clemmons LM (1995) "Nature was her lady's book": ladies' magazines, american indians and gender, 1820–1859. Am Periodical 5:40–58

Conkin PK (2002) Peabody college: from a frontier academy to the frontiers of teaching and learning. Vanderbilt University Press

De Jardins J (2010) The madam curie complex: the hidden history of women in science. The Feminist Press at CUNY

Driggers E (2018) Augusta Hasslock Kemp: Women, geology, and the West, vol 50. Presentation at The Geology Society of America Meeting, Indianapolis, Indiana. Paper No 200-9: Geological Society of America Abstracts with Programs, p 6. https://doi.org/10.1130/abs/2018AM-319897

England K, Boyer K (2009) Women's work: the feminization and shifting meanigns of clerical work. J Soc Hist 43(2):307–340

Goodman R (2014) How to be a Victorian. Penguin Books, London

Gries RR (2017) Anomalies: pioneering women in petroleum geology, 1914–2017. Jewel Publishing LLC, Lakewood

Hasslock AT (1910) Geology and geography of the area southwest of Abilene, Texas. MS Thesis, University of Chicago, Department of Geology. http://pi.lib.uchicago.edu/1001/cat/bib/4252742

Hazlett AJ (1921) First regional meeting of geologists calls out large attendance. Natl Pet News 14(4):25–26

Kemp AH (1965) Pegisis Limping. Davis Brothers

Kemp AH (1853–1974) Augusta Hasslock Kemp Papers. Texas Tech Southwest Collection, Lubbock. http://resources.swco.ttu.edu/guide/k.php. There were several other papers that were consulted there as well, and I wanted to include the guides. See footnote 8

Kemp AH (1882–1963) Augusta Hasslock Kemp Papers, 1617–1967. Tennessee State Archives, Nashville. https://tnsla.ent.sirsi.net/client/en_US/search/asset/20340/0

Kemp JF (1883–1943) John frank Kemp papers. Texas Tech Southwest Collection, Lubbock. http://resources.swco.ttu.edu/guide/k.php
Kemp AH (1934) Artificial oolites from spray from ammonia-cooling tower of ice plant, Seymore, Texas. J Sediment Res 4
Kemp AH (1957a) Color retention in stenopoceras, euomphalus, and naticopsis from the lower Permian of north central Texas. J Paleontol 31(5):974–976
Kemp AH (1957b) The siphuncles of some coiled nautiloids from the lower Permian of baylor county, north central Texas. J Paleontol 31(3):591–594
Kemp AH (1959) Pisolites formed from the oilfield water of the Luling field, Caldwell county, Texas. J Sediment Res 29
Kemp AH (1962) The stratigraphic and geographic distribution of cephalopod general in the lower Permian of baylor county, north Central Texas. J Paleontol 36(5):1124–1126
Larson EJ (2006) Summer for the gods: the scopes trial and america's consintituing debate over science and religion. Basic Books, New York
Light JS (1999) When computers were women. Technol Cult 40:455–483
Lightman B (2007) Victorian popularizers of science: designing nature for new audiences. University of Chicago Press, Chicago
MacKillop J (2016) A dictionary of celtic mythology. Oxford University Press, New York
Martin LD (1994) S. W. Williston and the exploration of the Niobrara chalk. Earth Sci Hist 13(2):138–142
McCoy D (1963) Saga of pebble puppies. Shared by Robbie Gries. The manuscript was transcribed through viewing Dorothy McCoy's Diary and Dorothy signed the document in 1963. Spank McCoy gave his permission to quote from the document and I thank Gries for sharing the document with me
Miller AK, Kemp AH (1947) A koninckioceras from the lower Permian of north-central Texas. J Paleontol 21(4):351–354
Moss LG (1996) Wild west shows: and the images of american indians: 1883–1933. University of New Mexico Press, Albuquerque
Pulliam JD (1968) History of education in america. Merrill Publishing, Columbus
Romer AS (1964) Augusta Hasslock Kemp. J Paleontol 38:1008
Rossiter MW (1980) "Women's work" in science, 1800–1910. Isis 71:381–398
Rupp LJ (1997) Worlds of women: the making of an international women's movement. Princeton University Press, Princeton
Russett C (1991) Sexual science: the Victorian construction of womanhood. Harvard University Press
Schiebinger L (2004) Nature's body: gender and the making of modern science. Rutgers University Press, New Brunswick
Second JA (2014) Visions of science: books and readers at the dawn of the Victorian age. University of Chicago Press, Chicago
Sklaroff LR (2009) Black culture and the new deal: the quest for civil rights in the Rosevelt era. University of North Carolina Press
Stevenson D (2013) 1914–1918: the history of the first world war. Penguin, New York
Terzian SG (2006) "Science World," high school girls, and the prospect of scientific careers, 1957–1963. Hist Educ Q 46(1):73–99
University of Texas Alumni (1964) Alcalde. March
University of Texas Alumni (1965) Alcalde. June, p 37
Vance T (2001) Women in paleontology: Augusta Thekla Hasslock Kemp. Dallas Paleontologycial Society: The Fossil Record (Dec 2021): https://dallaspaleo.org/resources/Documents/DPS%20Fossil%20Record%202021-12.pdf
Wagner TS (2016) Victorian narratives of failed emigration: settlers, returnees, and nineteenth-century literature in english. Routledge, New York
Ward I (2014) Sex, crime, and literature in Victorian England. Bloomsbury, New York
Wichita Daily Times (1947) Man's world in oil field vanishes: five feminine geologists listed here. 2 March

Chapter 3
"Women Will Win" Geology and Careers for Women in the World Wars

3.1 Introduction: Outlook and Openness

In 1916, the pages of the *Gazette Times* ran an ad which promoted the quality of woman with men. The Woman's Federal Oil Company of America wanted to provide opportunities for women in the oil fields ranging from Kansas, Oklahoma, to Louisiana. The company billed itself as successful, "The operations of this company are progressing in proven territory which has been producing oil and gas for more than twelve years, and on which full exact scientific reports by the United States Geological Survey are available" (The Gazette Times 1916). The company, wholly owned by women, was run on "science" and "the cleanest business principles," and all of the stock holders were women.

The advertisement was proudest about women being triumphant regarding the history of women in the oil and gas business. Woman were doing dangerous and challenging professional work,

> Now comes a group of sixteen individually successful women from different parts of the country, joining hands to offer a remarkable business opportunity primary to women. They have an enterprise that is truly remarkable I never sense, since they have applied scientific principles to the oil business, which is commonly supposed to be very hazardous, and have adopted the vest methods known to all the large corporations in this field, proving conclusively that the that law of averages makes the oil business as safe as the insurance business. (The Gazette Times 1916)

The women setting up the company felt confident in not only their abilities, and the market, but also of a world that was open to woman working in gas and oil. Unfortunately, the company does not last and ultimately closes under a special financial deal in 1920.[1]

[1] There are not enough studies on the women-run oil companies of the early twentieth century. See https://aoghs.org/old-oil-stocks/womans-federal-oil-company-of-america/#:~:text=Although%20later%20derided%20for%20having,exploration%20companies%20failed%20than%20succeeded

E. A. Driggers, *Glamour and Geology*,
https://doi.org/10.1007/978-3-031-64525-9_3

This chapter explores the optimism and openness of the US workforce for female geologists. Wars largely pushed for the positive outlooks for women entering geology and oil, but both academics and industry professional embraced and encouraged women entering into geology. US labor groups further encouraged women, especially those who embraced their femininity, to enter into geology for employment in oil and gas. The US government, especially its bureaucratic arms that studied women's labor encouraged women to enter into geology. Many labor groups and industry groups promoted war time stories with phrases like women "To Take Places of Men" (National Petroleum News 1918). This phraseology of replacement will continue to emerge throughout the early twentieth century.

3.2 Geology and Women in World War I

In 1918, the US Public Information Committee published *War Work of Women in College* (Committee on Public Information 1918). The pamphlet informs the reader about how colleges contributed to the war by helping women accomplish their new wartime "accomplishments." This is done through answering around 150 questions that were collected by American colleges who participated in the US war effort. There were "war courses" offered at these colleges that prepared women to meet the needs of the war effort, such as classes in stenography or home nursing, which students undertook but female undergraduates did not receive college credit. However, regularly offered courses that were also important to the war effort, such as "wireless telegraphy" and "map making," were forged by the geology and physics departments, in colleges like Wellesley (Committee on Public Information 1918).

During the early twentieth century, there was a type of optimism for women's participation in geology that likely continued to flourish during World War I. For instance, in the recording of The International Congress of Women in 1899, Miss C. Raisin in the United Kingdom reported on the status and participation of women in geology in the United Kingdom. Raisin optimistically framed this report as progress: "I HAVE been asked to give some account of the progress made by women in one of the sciences to which I have given my life-that of Geology" (Raisin 1899). She noted that women were active in the profession but it is hard to categorize their work as professional as they often did not receive payment for their services. But she believed that progress in science would have not occurred without the work of women: "But it would be anomalous if this subject were omitted in any account of women's progress in Science, for it is a subject in which the advance and increase of women's work in recent years has been most marked" (Raisin 1899). For instance, the Geological Society of London has received more papers from women that often got positive attention. The Society also gave its most prestigious award, according

as well as https://oilwomanmagazine.com/article/female-forces-a-timeline-of-history-making-women-in-energy/

to its benefactor Charles Lyell, to either men or women. Women were contributing to the creation of knowledge in geology but Raisin also inquired about how much contributions women were making to the more practical aspects of geology, such as the obtaining of natural resources. She argued that women are just as qualified as men to obtain resources like minerals and work on projects involving the "water supply" (Raisin 1899). But Raisin was encouraging of women contributing to knowledge creation.

In 1918, stories about women working in the oil field are posted in publications like Alumni magazines (Wellesley Alumnae 1918). In *The Wellesley Alumnae Quarterly* there is a an article about Elizabeth Fisher, who worked in the Geology and Geography department. She was called a "pioneer" as she was working out in the oil field: "There have been women geologists in office work, but Professor Fisher is a pioneer in field work." She spent most of her summer working in oil fields in the Texas area, especially the "north central" area that had emerged as a new oil and successful oil field (Wellesley Alumnae 1918). The article makes mention of the war emergency, which produced an oil shortage. The article praised that, "This past summer, when the oil shortage threatened our government so seriously, Miss Fisher was sent by a large Kansas oil company to the newest oil field in America, in north central Texas" (Wellesley Alumnae 1918). Oil had been produced in Texas in 1917 which was producing around 1350 oil barrels per day, and by 1918, it was producing 5000 barrels a day. Elizabeth Fisher found more useful oil wells: "…Professor Fisher was sent to find and to study those geological structures which indicate possible pools of oil underneath" (Wellesley Alumnae 1918). The article praised that in previously arid farmland, oil was producing wealth. Fisher also went to the International Geological Congress and traveled to Russia in 1897. She studied the oil fields while she was in Russia.

In 1920, Catherine Filene Shouse published an edited volume of essays regarding potential careers for women. One of the careers profiled is "The Geologist" (Bliss 1920). Eleanora F. Bliss of the US Geological Survey defines and describes the work of the geologists right after World War I. She describes a geologist with male pronouns, even though she is writing for a principally female audience. The description read as "The geologist is a person who studies the constitution and structure of the earth" (Bliss 1920). Interestingly, she also defines, "He observes the physical forces that are operating to produce the earth as we know it, and from these observations he draws conclusions as to the history of the earth's development in the past and makes certain predictions as to the probably course of its development in the future" (Bliss 1920, pg. 414). Bliss continually uses "he" to describe the work of geologists in a publication for women.

Important to the training of a geologist, according to Bliss, is performing field work and working with fossils. Much of the job of the geologist is working in the field and performing analysis and lab work inside. However, there were more office-based jobs Bliss emphasized for women: "There is office work for the geologist who does not wish to go into the field, and prefers to work over the material collected by others" (Bliss 1920, pg. 414). She noted that paleontology would be good for those persons interested in office work: "Paleontology is a particularly good field for office work and several women have already made notable success in this line of work." But in order to be a successful geologist, the careerist needed to work both in field and in laboratory and other office-based pursuits.

The physical challenges of the geologist, Bliss warns, will be one of the greatest challenges to women. The physical hardships of the job would test most women:

> The life of the field geologist is arduous and requires, in addition to an impelling enthusiasm for the subject, a physical strength and energy and an ability to endure hardships of a certain sort which comparatively few women as yet possess. It has many drawbacks at present as a vocation for women. (Bliss 1920, pg. 415)

However, women can make a living performing the work. And even if women do not want to pursue field work, there are several opportunities for more indoor work, such as bibliographical compilations of minerals and statistical analysis. The World War created new opportunities for women in geology and opened up new ways of contribution. One effect of the war was stress on the mineral reserves. The government realized that a close minoring of mineral resources was crucially important. The US Geological Survey took on the role of accounting for these minerals. And women could contribute to this important national interest: "Several prominent mining engineer offices throughout the country are maintaining information branches for the benefit of their clients, and in a least on case a woman is successfully conducing this branch of their work" (Bliss 1920, pg. 415). Women could also participate in the field of geology by becoming teachers.

There were a lot of places where women could get training geology. Bliss included Bryn Mawr, Vassar, Wellesley, Mount Holyoke, Smith, Barnard, Radcliffe, Cornell, and other universities, including other state universities. Those women wanting to go into statistical and bibliographical work would require, according to Bliss, a couple of years of college, but she encouraged that those women wanting to go into more research-based geology need to pursue more advanced degrees or graduate work and should take as many science classes as possible.

Teaching is likely a good way for women to advance in the field. Bliss saw teaching position for women in secondary schools and colleges. However, it is unclear about how women are going to advance in more commercially oriented geology. Bliss writes, "The chances for advancement in commercial work are still unestablished, as this is a new field for women. They must of necessity be largely dependent upon the individual, as the advancement of any individual in a business firm is conditioned entirely by the value of that individual to the business" (Bliss 1920, pg. 416). She cautioned that women working in mining often do not advance without field work, that she often cannot get. She emphasized that the US Geological Survey or State Geological Surveys are increasing the opportunities for women. Bliss estimates that more research-based jobs would provide a living of 800–4000 dollars in educational (teaching) geology. However, in the federal government, women working in geology would make 1200 dollars with a potential 240 bonus. Commercial-based geology would offer salaries of around 1800–2000 dollars.

In discussing qualifications for success, Bliss mentioned again that field work will be a challenge for women. Bliss reminds the reader that, "A woman particularly requires adaptability and poise, as her life in the field will throw her in contact with all sorts and conditions of men." (Bliss 1920, pg. 417) She also adds that: "Personal courage and well-controlled nervous system are essential requisites." (Bliss 1920, pg. 417) Bliss, in discussing the advantages and disadvantages, cautioned that women are severely hampered in the field, as she describes the difficulties of field work for women:

> It is hard to detail one woman to a camping party of three or four men. In order to live the life of a field geologist a woman must be strong and of an active disposition. The hours are long and a good field day covers from seven in the morning to six at night. (Bliss 1920, pg. 418)

She also adds that time cataloging samples and specimens taken make days add up to fourteen hours, accompanied by bad food and long physical exertion. Office work can also be long, without Saturdays off in some cases and a lack of time for summer vacation.

Bliss estimated in 1920 that the demand for female geologists is higher than the supply. She estimated, "The demand for geologists to fill office positions such as described under the Foreign Mineral Section of the Geological Survey and in commercial offices of oil companies and mining engineers will probably increase, and women will doubtless find in these positions assured and reasonably lucrative openings for geological work." (Bliss 1920, pg. 418)

By 1930s and 1940s, female geologists, especially women graduating college, saw lots of opportunities for women entering into petroleum. Some of those careers were driven by wartime need and war time success of women working in the field. Other opportunities were driven by the image of the female geologists, who was both feminize and professionally educated. The outlook for female geologists was becoming highly optimistic for women who were willing to fit the image and expectation of the working woman geologist.

3.3 The Outlook for Women in Geology During the World War II Era

Frida S. Miller, the Director of the US Department of Labor, Women's Bureau, wrote of her joy of transmitting a pamphlet about the prospects of women in the fields of geology, geography, and meteorology. She was excited to share the results of a study for careers for women in these fields in writing to the Secretary of Labor, the Honorable L. B. Schwellenbach, who served under Harry S. Truman. She wrote with optimism that, "…I have the honor of transmitting a description of the outlook for women in geology, geography, and meteorology which has been prepared as part of a study on the outlook for women in science. The extraordinary demand for women with scientific training during World War II and the resulting questions which came to the Women's Bureau prompted us to undertake this study"[2] (Schwellenbach and Miller 1948, pg. 7-III). Miller noted that there was a lack of resources like the one that she transmitted to the Secretary of Labor regarding women and science.

The pamphlet includes a list of several other similar resources for women in science broken down into different subjects: chemistry, biological sciences, mathematics and statistics, architecture and engineering, physics and astronomy, and "occupations related to science" (Schwellenbach and Miller 1948, pg. 7-II). Miller wrote that there was a lack of similar publication and that other female scientists encouraged her to publish other pamphlets: "The paucity of published information on women in science and the encouragement of the scientist and educators who were consulted in the course of this study confirm the need for the information here assembled and synthesized" (Schwellenbach and Miller 1948, pg. 7-III). Elsie Katcher was the author of the pamphlet focusing on geology, geography, and metrology.

The pamphlet was aimed at filling a need for getting women into science. The pamphlet promotes, "Much has been written about science and scientists, but little has been told about the work women trained science have done and can do in the future" (Schwellenbach and Miller 1948, pg. 7-V). The pamphlet highlights the contributions of women in science during World War II, but there are a few amount of women in science: "Although these women are few in number when compared to men in science or to women in such occupations as teaching and nursing, their contribution to the national welfare, so strikingly demonstrated in World War II, goes forward daily in the laboratories, classrooms, offices, and plants in which they work" (Schwellenbach and Miller 1948, pg. 7-V). It is the work of this pamphlet to point other women and new generations of women to scientific work. The pamphlet proudly proclaims its mission: "The every-day story of where these women work, or what kind of work they are doing, and of what other young women who join their ranks in the future may do has been the subject of this report on the outlook of women in science" (Schwellenbach and Miller 1948, pg. 7-V). The pamphlet looks at the trends of women entering the scientific workforce. It used June King as a representative model for a working woman in geology, see Fig. 3.1.

There were about eight hundred pieces of literature analyzed in making the pamphlet, but a lot of information came from industry, male and female scientists, and other employers. Many groups of female scientists provide information or were interviewed in constructing this pamphlet. Some groups that contributed to the construction of the pamphlet include the National Research Council, Office of Scientific Personnel, the US Bureau of Labor Statistics, the National Roster of Scientific and Specialized Personnel, the US Office of Education, the US Civil Service Commission, and the US Public Health Service (Schwellenbach and Miller 1948, pg. 7-V). The WAC (Women's Auxiliary Corps), WAVES (Women Accepted for

[2] Schwellenbach and Miller (1948) includes section numbers and page numbers. Page and section numbers will be given where available in citations.

Fig. 3.1 Image from *Outlook for Women in Geology*. (This image is likely from *Glamour and Geology*)

Volunteer Service), Marines participants were also consulted for the pamphlet. They interviewed women in industry and in some cases, "…women in scientific work were interviewed on the job" (Schwellenbach and Miller 1948, pg. 7-V; Yellen 2004). Women at Colleges and Universities were also consulted in regard to collecting statistics and other information about women in science.

The pamphlet then tries to define geology and the work that geologist perform. Geology and geologists are defined as: "Geologists study the constitution, structure, and history of the earth as it is disclosed by the sequence of formations and deformations or rock layers, and by the fossil or mineral content of such layers" (Schwellenbach and Miller 1948, pg. 7-X). In addition to the traditional geologist, the pamphlet also includes a definition of geologist who work more in industry: "The economic geologist deals with the exploration, exploitation, and study of useful mineral deposits, and the study of sties for dams and foundations" (Schwellenbach and Miller 1948, pg. 7-X).

At the time of the creation in 1948, the pamphlet notes that there not very many geologists overall and that women made up a small number. Geology, as a discipline, was just starting to embrace Ph.D. as a necessary degree. In historicizing the number of Ph.D. that were awarded, the writers of the pamphlet make the historical note that there were only 17 Ph.D. awarded in geology by 1944 compared to 55 in total prior to World War II. Most forecasters believed that there would be around one hundred Ph.D awarded annually by 1945 and then exploding to around four hundred in 1955.

The pamphlet points out that there might not be enough geologists to fit the needs of the future. However, it also notes that women received better opportunities because of the war, but the pamphlet did not know if this trend would continue. The pamphlet conveyed worries about how returning male veterans to work would affect the field of geology:

> In this even the women geologists, especially those with graduate training will continue to enjoy relatively better opportunity than they did in the days before the war. However, with men geologists returning from military service, and with the drop-off in demand peculiar to the war, women geologists already faced much greater competition in 1947 than during the war. However, since the supply of women students majoring in geology dropped during the war, it is unlikely that there will be an oversupply of women (Schwellenbach and Miller 1948 7-7, pg. 67).

The number of overall female geologists remained small. The female group (American Women of Geology) in the Geology Society of America (a major organizing group of geologists practicing in America) noted that there were about 271 female geologists, which was equal to about three percent of the overall number of geologists working in America. The fields that female geologists went into represented a notable shift when the American Women of Geology examined what subfields female geologists entered into during their working lives. Two hundred and three of the two hundred and seventy-one geologists were employed in 1945–1946. Most of the women employed were working in colleges and universities, including those working in research. Female geologists were also highly involved in research linked to the war. The number of women teaching geology improved compared to the time prior to the war. Working women geologists were also involved in industry, with about a quarter of the women surveyed by the women's subgroup in the Geological Society of America worked in the petroleum industry (Gries 2017). In comparing male employment, the petroleum industry made up half of the men working in geology.

Another quarter of working women in geology could be found in both federal and state agencies involved with geological surveys. The pamphlet notes that during the war the number of women working for state and federal governments was as high as forty percent. Teaching was forecasted to be the most consistent and active area of future employment for working women geologists: "Teaching at the college level, for example, will probably continue to be the most favorable field of women trained in geology" (Schwellenbach and Miller 1948, pg. 7–9). Women teaching geology, the pamphlet noted, also would be called on to teaching geography and any other related sciences. At schools and universities, women could also continue to involve themselves in research. However, of the number of women found to be teaching in universities in 1947 (40), the pamphlet noted that only 25 of these women were able to fully concentrate on teaching geology; they often had to teach other subjects. Other female geologists, particularly those in the petroleum field, were able to concentrate more on research, as well as those in graduate programs. It is also noted that of the forty women that were involved in higher education, only nine of those had "professorial" positions.

At the time of the creation of the pamphlet, it was found that around 188 women were involved in teaching geology at colleges and universities, only half were able to focus exclusively on geology, while the other half combined geology with some other subject like geography. However, most of the women were not working at exclusively female institutions, like women's colleges and universities. The pamphlet historicizes that fact, "Although in the past it was easier for women to secure teaching appointments in women's colleges than in coeducational institutions, the count of women geology teachers made in 1947 indicated that more women teachers were employed in publicly and privately controlled universities than in colleges of liberal arts and science, among which the women's colleges are usually classified" (Schwellenbach and Miller 1948, pg. 7–9).

The study noted that much of the money that was coming into universities, in addition to "private research projects," encouraged women to work in geology and motivated them to pursue their geologic education ended after World War II. Most women lost their jobs that were involved with research and the war effort, "Practically all women were released after VJ-day when these research agencies began to return to their peacetime size" (Schwellenbach and Miller 1948). Women often lost their job because they did not meet the technical requirements needed to remain on the project full time, "Few women could meet the highly specialized requirements set for permanent staff members" (Schwellenbach and Miller 1948). But research that was discontinued prior to the war might bring about a return of research positions for those same women, "However, with the revival of some of the normal research activities interrupted by the war, there is some opportunity for women as research assistants and as library and editorial assistants in research agencies and publishing houses" (Schwellenbach and Miller 1948, pg. 7–10). Women could also find employment in other related areas, such as jobs that dealt with public engagement, "Popular geological writing for travel agencies, nature clubs, magazines, and children's books offer another outlet for women trained in geology who have writing ability" (Schwellenbach and Miller 1948). In summary, as many opportunities that the war provided in regard to research, those opportunities were not projected to return in the same ways.

Women had a more optimistic outlook in regard to entering or maintaining their position in the oil and gas industry. The pamphlet boasted, "A year after VJ-Day [Victory over Japan], women seeking jobs in the petroleum industry found that the attitudes of industrial employers had definitely been tempered by the war" (Schwellenbach and Miller 1948). The pamphlet promoted women's exceptional work in the laboratory. However, the pamphlet pessimistically noted that women in the petroleum industry were encouraged to take more clerical and office-based jobs, or other jobs that were considered para-professional. Though women had the ability, they were not being offered jobs in more technical capacities, "There was greater recognition of the ability of women laboratory technologists in research departments, but by and large the opportunities open to women were those which had been open (in few numbers) before the war. Women with backgrounds in geology were encouraged to take clerical or drafting jobs, where they might be especially valuable in assisting staff members in their work" (Schwellenbach and Miller 1948, pg. 7–10).

Women who were studying for graduate degrees in related geological fields were encouraged to remain working in technical jobs in the oil industry. The pamphlet points to two women who were working in jobs that focused on geology at the microscopic level. These women examined rocks and other geologic samples with microscopes: "…most of the women with graduate training in geology who were working in the oil industry in 1947 were engaged in the microscopic examination of rock cuttings and cores from well borings, studying the lithology and faunal and floral contents of the rocks" (Schwellenbach and Miller 1948, pg. 7–10). The pamphlet writer juxtaposed the work in geology: that in the micro and the macro. The writer notes that women were "well suited" for the work of micropaleontologists: "Many companies have found that women are well suited for this work not only in routine analysis but also in research" (Schwellenbach and Miller 1948, pg. 7–10). The micropaleontological work is pitched as advantageous for women at the corporate level: "Many companies have found that women are well suited for this work not only in routine analysis but also in research." Women are also recommended to pursue positions as "petrographers, stratigraphers, and geophysicist," which the pamphlet pointed out required advanced degrees.

The pamphlet implied that field work was more of a possibility for women in the post-World War II era, but they had to demonstrate their fitness in ways that men did not have to during that same period. Reflecting the idea that women had to prove themselves worthy of performing field work, the pamphlet pointed out what the writer thought was an opportunity: "Women who can prove their abilities in these specialties are sometimes sent into the field when their work requires it" (Schwellenbach and Miller 1948, pg 7–10). The pamphlet notes that two women were working in oil companies abroad, such as in South America and even Cuba.

By 1947, the pamphlet notes that there were only 25 female geologists working in the US Geologic Survey, with two of its geologists working in Japan at that time. Because of the war efforts, the pamphlet points out that around 50 women were working for that organization. In order for women to serve as geologists in the government, they had to pass a civil service exam for geologists. The need for geologists after the war was so great that veterans and those geologists working for the government would get preferential treatment: "The register established from this examination was not expected to be large enough to fill all the vacancies that existed. Under civil service regulations, however, preference is given to veterans and to former Government employees with civil service status" (Schwellenbach and Miller 1948, pg 7–11).

The pamphlet laments that men will continue to get preferential treatment in the post-War workforce. The writer noted that while geologic surveys both at local and at national levels would return to monitoring water and more routine geologic work and collect field data: "Undoubtedly, the preference for women in these positions will continue, but the woman geologist who is a good as or better than a competing male geologist will have some opportunity for employment" (Schwellenbach and Miller 1948, pg 7–11). Unfortunately, women would only get a slight advantage in the geologic job market if they were equal or superior candidates to male geologists.

Women were encouraged in the pamphlet to look for jobs in the geographically advantages areas that produced oil. Those areas of encouragement included the United States South, Washington, D.C., and areas of the Eastern part of the United States near colleges and universities (7–11). Women had to be able to move their lives around compared to other professions like teaching and nursing. There were also industrial geology jobs in areas like "Texas, California, Oklahoma, Louisiana, Kansas, and Illinois" (Schwellenbach and Miller 1948, pg. 49). These states prior to World War II produced most of the oil for the United States. Most of the women listed in the American Women of Geologists were involved with the federal government, while the rest were employed in the North-Eastern Part of the United States and teaching geology in colleges and universities. Other female geologists were involved in geological surveys, research, and industrial geology in mining and oil jobs.

The pamphlet goes on to recommend how women can pursue a career in geology. It reminds those women and girls pursuing geology that they had to be healthy and "physical stamina to enter their field" (Schwellenbach and Miller 1948, pg. 7–11). Girls should study drafting, as well as studying typing, and the practice of taking formal notes (stenography). They also needed to prepare to write clearly, as this would be required when they were composing formal reports. They were also encouraged to study a foreign language as well. Training in math, chemistry, as well as physics was strongly recommended.

Taking classes in botany and biology was also recommended, which were usually covered in undergraduate degree courses. And most of the major fields of geology were encouraged: "micropaleontology, economic geology, stratigraphy, petroleum geology, and sedimentation, before selecting a specialized field for graduate study" Schwellenbach and Miller 1948, pg. 7–12). The pamphlet also encouraged five years of college training. But by 1946, only half of all geologists had an undergraduate degree. However, when the writer examined the geology field, they found that more women had a masters, but more men had doctorates. Conveying the post-war forecasting of the job market stated that geology was going to require a Ph.D., and the writer of the pamphlet noted that it was "doubly significant for women" to pursue a doctorate.

3.4 Training a Female Geologist

Caroline E. Heminway worked at Smith College as a geologist (Kierstead 1923–1985). She was a curious student, who pursued philosophy and psychology, with a geology minor at Mont Holyoke College, located in Massachusetts. She later focused on geology exclusively and received a master's degree in geology from Cornell, and after that received her Ph.D. at the University of Indiana. She later married Friend H. Kierstead, changing her name to Caroline Heminway Kierstead by 1947. She worked at Smith teaching geology from 1928 to 1969, eventually rising to the rank of full professor, after a brief stint working at Shell. She also had success breading Corgis.

Heminway was a geologist who published, along with her colleague J. J. Galloway, a survey of *The Tertiary Foraminifera of Port Rico* in 1941 (Galloway 1941). As an established geologist, she was called to comment on the proper methods of training female geologists in 1947.[3] She published an expansive explanation and roadmap of how to train women in geology in the interim proceedings of the Geological Society of America. She noted that the training of women in geology is an important question that has not received enough attention in the profession. She sets up her discussion by highlighting the importance of women in geology:

> The question of the training of women in geology is one that has received considerable attention from those few of us who have been concerned with women in the geology department of the various women's colleges and coeducational institutions of this country. But, in view of the relatively small number of women who are actively engaged in geological work, this question has undoubtedly received only scant attention from the profession as a whole. (Heminway 1947, pg. 66)

There were women in the field, Heminway pointed out, and their numbers were going to increase. She pointed out that the World War had made women enter into geology, but not at the same rate as there was demand for women in the other physical sciences like math, physics, and also chemistry to serve the war efforts: "Undoubtedly, the war emergency had some influence on this increase in numbers, although there was no such strong and widespread demand for women geologists as there was for women wo were training in mathematics, physics, and chemistry" (Heminway 1947, pg. 66). She pointed out that there were 8300 geologists in 1945, of whom 256 were female, and half of those women were active in geology, while the rest were either working outside of geology or not working because of marriage or retirement (Heminway 1947, pg. 66). Heminway pointed out that the oil industry occupied most of the women working in geology and geology-related fields. However, teaching in colleges and universities accounted most of the remaining women working in geology.

Geologic organizations, Heminway pointed out, were changing language in their mission statements to include women. For instance, the American Association of Petroleum Geologists changed their constitutional statement on membership dating to March 1917 from "any man who met the qualifications" to "any person" in 1918, but women did not join the group until 1919 (Heminway 1947, pg. 67). Prior to World War II, the number of women in the American Association of Petroleum Geologists (AAPG) did not rise rapidly, but around the time of the war, the group saw an expansion of female membership. Heminway characterized the growth rate of the AAPG at around three percent, but that growth rate changes more rapidly, to having sixty-one members in 1946. Requirements in AAPG include field experience and an undergraduate degree, which according to Heminway: "...would seem to suggest that an increasingly large number of young women are going into

[3] Cited in *Outlook for women in geology, geography and meteorology* on pg 7-45.

commercial oil-company work either directly after the completion of their undergraduate work or after a very short period of graduate study" (Heminway 1947, pg. 67). Historical information from Smith College led Heminway to believe that undergraduate students studying geology seemed to follow a trend of industrial and educational career paths. However, she also notes that marriage did take some of those graduates out of career paths in geology (Heminway 1947, pg. 67).

Heminway strongly advocated for the same training of female geologists as their male counterparts. It will help, according to Heminway, the field of geology as a whole: "As Dr. Croneis has pointed out from time to time, our profession receives all too little publicity, and the average layman is very ignorant of geological facts. These nonprofessional women will help to improve the general level of popular knowledge of earth sciences throughout the United States" (Heminway 1947, pg. 67). Carey Cronies was a curator of museums at the University of Chicago and an active geologists both at state level surveys and in promoting the field of geology (Heminway 1947, pg. 67).[4] It is interesting that Heminway divides female geologists between nonprofessional and professional as it is unclear which characteristics contribute to the dichotomy that she continually refers to in the article.[5]

Heminway cautions against training for undergraduate women that is too technical, as it is helpful both on the job market or in applying to graduate programs (Heminway 1947, pg. 67–68). Students should focus on specialization in their graduate programs. One thing that seems to characterize the field, according to Heminway, is that it is unclear what makes a geologist a geologist, or when can one call themselves a geologist. Heminway proposed that the senior seminar should be the key indicator:

> The senior seminar should be a synoptic course, preferably taught from a regional point of view. This course should emphasize individual work by the student and should correlate and bring together as a coherent whole the carious fields of geology that will have been presented to the student up to the time as more or less separate entities. If all geologists, not merely women, were required to take such a course, there would be more unity among geologists and less acrimonious discussion on such topics as, "When is a geologist a geologist? Is a paleontologist is a geologist? [O]r, Is a physiographer a geologist?" (Heminway 1947, pg. 68)

Heminway emphasized the importance of the senior seminar for female geologists, as it would show them how interrelated and involved all of the fields of geology are to each other. It should also teach the female candidate (as she continually refers to the student as she) how to research by using bibliographies to "…run down source materials of various kinds, evaluate diverse and properly documented reports" (Heminway 1947, pg. 68). They should also be able to write their own clear reports. Pairing good writing with drafting sets up a woman for success: "Training of this kind, together with the training in drafting which was advocated above, seems to be the two types of somewhat specialized yet basic training that a woman

[4] See the biographical entry for Cronies at *Handbook of Texas Online*: https://www.tshaonline.org/handbook/entries/croneis-carey-gardiner

[5] For an introduction to the professionalization of geology historiography, see Porter 1978.

can be most certain of using, whatever her exact position in geology may be" (Heminway 1947, pg. 68). The senior seminar helps a candidate, but Heminway also recommends at the end of the program, as she provides a list of suggested courses, "One specialized course, two semesters, 6 hours of credit" (Heminway 1947, pg. 67). Heminway recommend this focus in the advanced subfields of geology: "Advanced minerology with optical mineralogy and petrology, Micropaleontology, with the chief emphasis on Foraminifera, Economic geology, Stratigraphy, Petroleum geology, or Sedimentation" (Heminway 1947, pg. 69).

The emphasis on the capstone course for students was important because it set them up for success after they left school. Heminway summarized the advantages: it provided specialized training that would allow the student to determine if they wanted to pursue more advanced education, it set them up to succeed in industry or doing additional education, and it makes women better geologists. Heminway wrote that women did not get a chance to go to graduate school very often or have many opportunities to pursue advanced training. She lamented: "(3) Since many women neither go on to graduate school nor become professional geologists, this specialized course strengthens and enrichens their knowledge of geology" (Heminway 1947, pg. 69). Therefore, the senior capstone course becomes an important opportunity for female undergraduates receive advance training since it is rare that they are able to after their undergraduate training.

She also put emphasis on field experiences and field training for female geologists, as it was also important to male geologists. She emphasizes the importance of field work for women and geologists in general: "The course in field geology probably needs no justification, even for a woman. Although women are rarely employed as field geologists, they need this training to make them appreciate and grasp more fully the other phases of geology with which they may be more immediately concerned. When students, men or women, lack this field experience, I always fear a reaction such as was reported to me some years ago" (Heminway 1947, pg. 69). Heminway then related a story about a student who had little to no field experience. She wrote of a female undergraduate who had done well in her first geology class, as she received an A in the course. As the undergraduate was talking to her sister, her sister inquired as to whether she was going to take more geology courses. The sister responded no as she did not think that the course was accurate or "true" (Heminway 1947, pg. 69). In an attempt to prevent these experiences, Heminway emphasized the importance on fieldwork: "I am always fearful that, without adequate training and experience in the field, there will be more students who 'don't believe a word of it'" (Heminway 1947, pg. 69).

Heminway also emphasized the mathematics portion of geology education. She should pursue mathematics training beyond the first year and also pursue more laboratory sciences, in addition to foreign language. Heminway ended her talk with the final point that a female geologist needs to have somewhere between four and seven years of professional training. But she was confident that the undergraduate education that women received would set them up for proper graduate training.

The session that Heminway participated in was likely chaired by Dr. Mather. The Dr. Mather referred to in the Interim proceedings that led to the Question and

Answer portion of the talking that Heming gave about how to train female geologists was Dr. Kirtley Fletcher Mather, who was involved with the "Scopes Monkey Trial." Mather encouraged the audience to both identify themselves for the record and ask questions to Dr. Heminway. Mather praised her presentation, emphasizing its value to the field: "I am sure that all of us would heartily agree upon the first comment. It was an excellent presentation of some very interesting and important facts, and we do believe all the words of it, Miss Heminway. It was very well done. The meeting is now open for discussion" (Heminway 1947, pg. 70). Interestingly, Mather addresses her as Miss, even though she received her Ph.D. from Indiana in 1941 (Smith finding Aid). Also Mather was referring back to a previous quotation where Heminway said that some female undergraduates do not believe the information in their geology courses without solid field work experiences.

Heminway jumped into the after session by wanting to make a clarifying point. She wanted to clarify why she emphasized the high number of 39 semester hours in geology. She gave that high a number because of women not professionalizing prior to leaving college and that high amount of course work would be their last chance. She also compared and contrasted the situation that female geologists found themselves in compared to their male graduates: "I think that, in view of the fact that we have some women who do not become professionals, and for whom their college training is the last training they get in geology, we should probably expect a woman to take more geology as an undergraduate than we would a man" (Heminway 1947, pg. 70). Also Heminway added that women went into industry, even for a short period of time, prior to marriage, and she thought that they needed proper training to prepare them for their quick entry into industry. Heminway thought that this was a good thing.

Arthur Bevan, a University of Chicago–trained geologist, secretary of the American Geological Association, and head of the Illinois Geological Survey asked if Heminway had thought of women working as secretaries in industry, as they would have robust technical knowledge about geology (Leighton 1968). He added, "Of course there are quite a number of women there, but it seems to me there is quite an opportunity in this work for women in high clerical capacities or as crackerjack stenographers who really know the language of geologists"[6] (Heminway 1947, pg. 70). Heminway responded to the question with an emphasis in the training of undergraduate women and from her own experience.

Heminway thought that there was an opportunity for female geology majors to fill those roles, but she said it would be difficult to convince a graduate to take such as position. She responded: "From my own experience, I do not know many women who are in that capacity and I myself can report the reaction of the average undergraduate. She considers such a position very much beneath her when she has bothered to take this training, and I do agree with her" (Heminway 1947, pg. 70). However, Heminway noted that any graduate could do such a secretarial job with

[6]The conversation about women in geology continued in the same publication that included Heminway's talk. The questions and comments from the conference were included in the Interim proceedings.

about a year's worth of experience in geology. However, she emphasized that most geology graduates would not be interested in such a secretarial position, considering their training:

> I think that an intelligent woman who has majored in any field is competent and able to pick up, in the period of perhaps a year, enough geology to do the sort of secretarial work that would be required of her without the actual training as a major in geology. And I can't think offhand of any of our graduates who have gone into that sort of work. They are certainly not interested. (Heminway 1947, pg. 70)

Perhaps the emphasis on the lack of interests punctuated Heminway's response. Dr. Bevan replies with a rebuttal of sorts, replying that graduates should demand higher wages, as they have special skills. He backpedaled stating, "You can't expect a graduate of Smith College or any other women's college, let us say an ordinary liberal arts college, to go in at prevailing salaries. My viewpoint is that they should be paid what they are worth. They are a special classification, and I think it would help raise the level of public service." There was no further discussion and another geologist was called on. Again, even in the transcript, Heminway is referred to as Miss Heminway, though she had a Ph.D. awarded in 1941, and all of the other commenters are referred to as "DR" (Heminway 1947, pg. 70).

The next question was asked by a "DR. Rich." This is presumably Dr. Richard L. Lougee, who was supposed to present the next session "Cultural Versus the Professional Approach to Geology for College Students." Dr. Rich asked Dr. Heminway, though he does not identify her as doctor, as to whether female geologists were working in the fields of map making or photography from airplanes. He thought that female graduates should be working in this field. The positives, he argued, included extensive travel. But he added that there might be drawbacks: "They should see quite a lot of the country. I suppose that is the main drawback in it, but very large parts of the earth are going to be studied from aerial photographs, and I don't see why a keen young woman shouldn't do that as well as anybody else, perhaps better than most, provided she has seen enough country" (Heminway 1947, pg. 70). Heminway agreed that women were doing that work during the war, but she did not know how many opportunities remained for women to enter into those position in the current moment. She ended her response by promoting the work of women with maps: "I agree with you that this would be a very good opening for women because they are supposed to be able to master techniques and work with maps in a very much shorter period of time than men" (Heminway 1947, pg. 71). Mather then transitioned the session to the next paper by making an awkward transition, he said that the next paper would be given by Doctor Richard J. Lougee but then added: "You will note that the transition from one speaker to the next should not suggest any implication that women may not be involved in the cultural approach to geology" (Heminway 1947, pg. 71).

Karl Ver Steeg published a response letter to the article, "Is Geology Easier for Boys Than for Girls?" (Steg 1933). Ver Steeg also examines the grades in his physical geology course and compared the grades of male and female students. He found that from his class of students, whose ages ranged from 17 to 21, with Ver Steeg noting that he taught mostly freshman and sophomores in regard to grade level. He noted that the course was mostly lecture based with a textbook and laboratory. In addition to the classroom- and book-based curriculum, there were field trips and the students were also encouraged to do additional readings outside of those mentioned in the class curricula. Most of the lab sections were about twenty five students which provided more interaction between the laboratory teacher and the student compared to the lecture hall. After analyzing the grades of 647 students, with 254 male students and 393 female students, most students were getting a C grade. He found that the women achieved slightly higher than the male students (Steg 1933).

Opportunities for women were starting to open up in geology because of the World Wars. Advertisers and car parts and manufactures wanted to reach the expanding market, while still promoting women working in geology as glamorous. Some car ads even targeted these working women.

3.5 "Lady Motorist:" The Working and Driving Woman

The petroleum industry made efforts in newspapers and journalist portrayed the gains of oil through the advance of automobiles, and they promoted those gains to women. Cars from the 1930s into World War II were highly promoted by the oil industry to women. The working woman geologist had cars advertised to her to make her work easier. Even at the current meetings of the Geologic Society of America (GSA) cars and car sales are prominently featured. With the rise in employed women during World War II and the post-war period, consumption of cars rose and manufacturers started to market directly toward women. Cars became a gendered technology but manufacturers wanted to sell them to working women (Nicholas 2024).

Some car parts manufacturers like Borg-Warner made ads that targeted the working female geologist in the oil industry. Working geologist Mrs. Phillips was identified as one of the few women working in geology and that she was fully engaged with working in the laboratory, with her microscope, at Sun Oil Company (Life 1950). Part of her job, according to the ad, was traveling out to the oil fields and taking trips, in addition to the recreational trips with her husband. These trips resulted in around 2500 miles a month onto the family car. But with the overdrive product produced by Borg-Warner, working women did not have to worry about the life of their car, as represented in Fig. 3.2.

The petroleum geologist Mrs. James T. Phillip then proclaims that the transmission helps her get her jobs done: "My work trips to the oil fields are so much easier with B-W Overdrive" (Life 1950).

Stories about women drivers also appeared in the literature produced by petroleum companies. There was a tacit acknowledgment that female drivers were on the rise, as the article reported that the number of women driving cars rose to 35 million, which was around double compared to the previous decade. Though the article opened with a sexist joke about female drivers not understanding how a car runs, the

Fig. 3.2 The quality overdrive transmission for working women. (Life 1950)

article admitted that "The statistics are impressive" (Klein 1961, pg. 27). This means that the opportunity to make money and market opportunity sometime outperforms practical sexism. This is expressed with the cartoon in Fig. 3.3, then contrasted with the glamour shot of the female motorist in Fig. 3.4.

However, *Petroleum Today* and the article's author Kerry Klein continued to frame the opportunities to profit off the expanding female driving population in misogynist terms. New service stations were designed to have the working and driving women feel welcomed and even employed the designer Frank Lloyd Wright. Filling stations designs focused on enticing women to stop and spend money: "The Dealer's first job is to coax the ladies to drive into this station, and thus the designer preparing plans for a new station has the spirit of America's female motorist standing at his elbow." The new filling stations were meant to be welcoming to what men perceived women wanted: "It is to attract and please her that he specifies extra-wide driveways so that she won't have to worry about maneuvering in a narrow space. It is largely for her that he includes walls of glass, bright lights, and spacious restrooms. Some of the newest stations even have a langue and television room" (Kelin 1961, pg. 27). Survey information had revealed that women wanted clean restrooms and that was the main factor in their plans to stop there. *Petroleum Today* encouraged car dealers to treat the female motorist with respect as customers for, "Once the lady motorist has driven into this station, the dealer must know how to please

Bob Thomson was just about to close his Gillette, Wisconsin, service station when two cars came up the driveway, one pushing the other.

"I don't know what's wrong," said the lady at the wheel of the first auto. Thomson checked and returned. "You're out of gas," he reported.

The lady sighed uncertainly, and a small frown creased her forehead. "Will I damage anything if I run it that way?" she asked.

Jim Parker operates a station in Cedar Rapids, Iowa. He likes to tell of the woman who wanted him to give her car a real going-over. She was worried. "Do a good job," she urged.

"Everything's fine," Parker announced after a few minutes, "only I had to put some water in the battery."

The woman nodded sagely. She looked up at Parker with an air of satisfaction. "I *thought* I heard something squeaking under the hood," she said.

Fig. 3.3 Sexism and female motorists. (Klein 1961)

her. She may not know much about the mechanical operations of her auto, but she knows when a station attended shows care for her car" (Klein 1961, pg. 28).

The article continued to express how dealers and service station owners could make their business more attractive to women, such as the employees being dressed in business suits and stressing professionalism. Again, referring to survey information about the preferences of women, Klein stressed, "Studies have demonstrated the fact that all customers, but particularly women, are impressed by a dealer who looks upon himself as a modern businessman. IT takes more than the oldtime grease monkey to run a successful modern station" (Klein 1961, pg. 28). The piece emphasized that women wanted to be treated with respect: "In the final analysis, though, it is the little added attention to milady's needs, the extra courtesy, that makes the sale."

The oil industry considered female drivers to be an important group of consumers and spent money with focus groups. Oil companies acknowledge that competitors were challenging them for customers, "The companies have used a wide variety of programs and techniques to encourage dealers to adapt themselves to the new importance of the lady customer. They have arrived at these rules of thumb by going directly to women autoists and asking questions." Answers they received stressed the desire of female consumers or "autoists" for value, proper service, the repairs being successful, and the car being cleaned prior to its return. Some car dealers took head of the article's emphasis that women wanted to buy cars with women present, and car dealers wives attended auto sales. And oil companies and car manufacturers were serious about making sure that service station and car dealers were attentive to female customers, "To help dealers check on their stations' appeal to women,

Fig. 3.4 Glamourized female motorist. (Klein 1961)

companies send around lady inspectors to suggest service or housekeeping improvements. One firm has a mystery lady who drops in unannounced and awards $1000 bonds to outstanding dealers. The local publicity resulting from the awards inevitably brings a boost in female business" (Klein 1961, pg. 28). Dealers, sellers, and proprietors of service station took efforts to attract women into buying cars and utilizing services, "And in a direct appeal to the women, several companies have lady representatives making the circuit of club meetings to discuss such subjects as automotive servicing, how to travel, what to take along, and how to pack it. Meanwhile, they are busily promoting their firms' gasoline and service" (Klein 1961, pg. 28).

Klein continued to push male proprietors and dealers to appeal to the female customer. He reported that dealers and operators were cleaning up their business, and oil companies would do what it took,

> Some oil companies and stations have gone to great lengths to capture milady's patronage. One group of stations gives women their change in "clean" money: the ladies have been found to be averse to dirty, greasy bills. A company prints its credit cards in pastel colors, these being more pleasing to the feminine eye. Mirrors in restrooms have come in for inspection: women don't like a marred looking-glass that distorts their appearance. A Connecticut dealer even sends flowers to lady customers who are in the hospital. (Klein 1961, pg. 28)

The article cautions that male dealers and gas station operators might want to ignore these changes to their business, the female consumer was a key component to their business, and they likely anticipated the business from female drivers to continue to grow. The article reminds the business operator that female consumers will have the last laugh if their business is ignored: "For however much the dealers may shake their heads over milady's whims and fancies, they are learning that the woman motorist is the key to a profitable operation. They may make all the jokes they wish, but milady is having the last laugh" (Klein 1961, pg. 28).

During the War years and into the post-War period, the US government saw geology as a real growth area for female employment. Car companies and industries related to the automotive industry wanted to market themselves to the growing numbers of female consumers who needed cars for their employment. Some of the ads, as well as employment pamphlets, emphasized the glamorous nature of the working female geologist. The next chapter will explore how geologist continued to be associated with glamour during and after World War II.

References

Bliss EF (1920) The geologist. In: Filene C (ed) Careers for women. Houghton Mifflin Company, New York

Committee on Public Information (1918) War work of women in college. Government Printing Office, Washington, D. C

Galloway JJ (1941) The Tertiary Foraminifera of Porto Rico. The New York academy of sciences. Scientific survey of porto rico and the virgin islands

Gries RR (2017) Anomalies: pioneering women in petroleum geology, 1914–2017. Jewel Publishing LLC, Lakewood

Heminway CE (1947) Training of women in geology. Interim proceedings of the geological society of america: 66–69. Cited in Outlook for Women in Geology, Geography and Meteorology, pp 7–45

Kierstead CH (1923–1985) Caroline Heminway Kierstead papers. Collection Identifier: CA-MS-00140. Northampton, Massachusetts: Smith College Archives. https://findingaids.smith.edu/repositories/4/resources/47

Klein J (1961) Madam motorist laughs last. Petroleum Today 2(3):26–28

Leighton MM (1968) Arthur Charles Bevan. https://www.stategeologists.org/sites/default/files/remembrance/Arthur-Charles-Bevan-1888%2D%2D1968.pdf

Life (1950) My work trips to the oil fields. p 50

National Petroleum News (1918) Employ women at refinery yard work. 7 August

Nichols NA (2024) Women behind the wheel: an unexpected and person history of the car. Pegasus Books, New York

Porter R (1978) Gentlemen and geology: the emergence of a scientific career, 1660–1920. The Historical Journal 21(4):809–836

Raisin C (1899) Women geologists. In: International congress of women in 1899: women in professions, vol 3. Ed. Ishabel Gordon Marchioness, pp 165–167

Schwellenbach LB, Miller FS (1948) The outlook for women in geology, geography, and meteorology. Bulletin of the Women's Bureau No. 223-7. U.S. Government Printing Office, Washington, D. C

The Gazette Times (1916) Women will win. 27 October

Ver Steg K (1933) Is geology easier for boys than girls? Science 77(1989):169

Wellesley Alumnae (1918) News from wellesley. The Wellesley alumnae quarterly. October

Yellen E (2004) Our mothers' war: american women at home and at the front during world war II. Simon & Schuster, New York

Chapter 4
World War II Glamour: The Image of the Female Geologist, and the Newspaper Pin-Up

4.1 Wartime Glamour

Appearing in *The American Weekly,* a wartime profile of June King was featured along the headline: "Mixing Glamour and Oil" in September of 1945. The one-page write-up was meant to convey the image of the feminine scientist, highlighting how the wartime American machine compressed female identity into a singular duality encompassing feminine virtues and technical/scientific expertise. To the American war effort, both went hand and hand, and were linked. Female geologists like King became celebrities in newspapers and publications that advertised the female geologist and her contributions to the war efforts (Fig. 4.1).

In 1938, *The Evening Republican*, published in Columbus, Indiana, published a story titled "Women Enter Geology." The subtitle was even more eye catching: "More and More Take Up Work., Some Being Employed in Oil Fields" (The Evening Republican 1938). This headline ran prior to the American entrance into World War II (1939–1945), though the United States did not enter the war until 1941. The field of geology, the story that originated from New York, empathized that more women were being attracted to the profession. The story implied to the reader that the author was shocked that women were transitioning into the field, "Geology, that practical science which appeals to the rugged, outdoor type of man, is attracting more and more women" (The Evening Republican 1938). The article quoted Ida Helen Ogilvie, who was the head of the geology department at Bernard, whose report was drawn from the presentation given at the Geological Society of America Meeting. The article quoted the major finding of Dr. Ogilvie's report, "The report, emphasizing that women have no physical handicap to overcome in geological work explains that women already are holding responsible jobs in coal and oil fields, as well as undergoing the hardships of scientific expeditionary work" (The Evening Republican 1938). Most women who are employed in the field of geology

E. A. Driggers, *Glamour and Geology*,
https://doi.org/10.1007/978-3-031-64525-9_4

Fig. 4.1 June King on the job (The American Weekly 1945)

are working in the oil industry, in offices and in research, with some of the female employees working in the field, drilling for oil.

4.2 "Mixing Glamour with Oil:" Female Geologists and Pin-Up Culture of World War II

The profile in *The American Weekly* went on to describe King as walking between two dichotomies that were thought to exist in the readers mind: "June King Is Demonstrating that IT's Possible to Be a Geologist and very Pretty, Too" (The American Weekly 1945). King was a stranger to the image of the oil field man: masculine and a "he-man." The oil field man was rigid and in the field, but women like King, were the new occupiers of that space: "A HARD-BITTEN Southern drill crew listened with magnificent unconcern a couple of years ago to word that a new geologist named J. King would be out soon to help search for oil" (The American Weekly 1945).

King was involved in field work, as the profile noted, that the foreman, might have described her as "'Jut another young squirt fresh from college,' growled the foreman, 'Well, we've broken in plenty of 'em. Let's get this rig movin'." The Carter Oil Company, which employed King, often listed her as J. King, and this led to surprise in her presence on the oilfield. The J, the profile empathized to keep the reader's attention: "…the J. in the name stood—not for Jack, Kim or Jake but—for a very feminine June" (The American Weekly 1945).

King had studied geology at the University of Kansas, and then her physical appearance was further emphasized by the profile as: "….a pretty brunette who

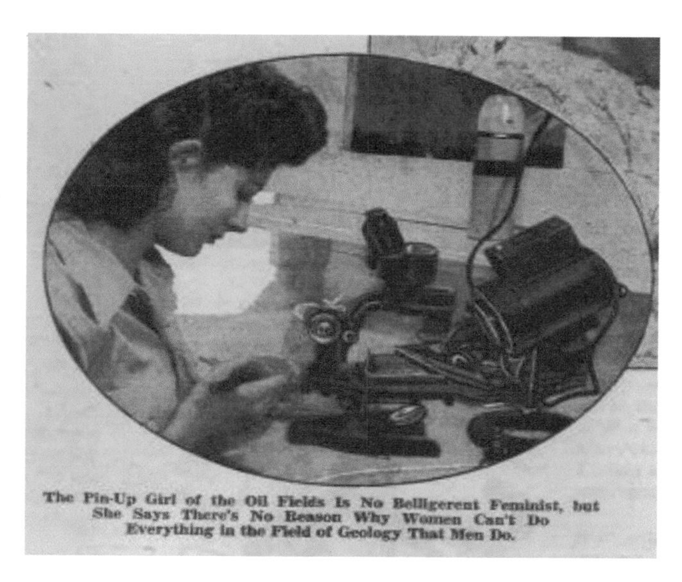

Fig. 4.2 King in the laboratory (The American Weekly 1945)

might be an actress if she hadn't turned to mixing glamour with oil." She was pictured with her microscope and laboratory equipment, as pictured in Fig. 4.2. The effects of her physical charm was described, "Several tough drillmen almost fell over themselves trying to help her, but the girl geologist quickly showed that she knew how to handle and chart the rock cuttings brought up as samples of the strata far below the topo f the ground" (The American Weekly 1945).

The older drill workers did not trust her at first, or enjoy a woman on the job, and inquired if she could truly make additions to the oil exploration that was going on in their fields. After citing her scientific expertise, the article conveyed the concerns about her physical appearance: "June was nice to look at, they conceded, but they questioned whether a woman—especially such a disturbingly pretty one—could make any serious contribution to the oil industry" (The American Weekly 1945). Her confidence quickly shut down the detractors. The article conveyed her confident actions and her abilities to change attitudes: "But June's businesslike manner soon silenced the doubters. Probably she's aware of her charms, like all comely girls, but during the working hours she's a geologist first. She tramps over the drilling fields in dungarees, workshirt and heavy boots and doesn't mind getting muddy" (The American Weekly 1945). She also had a strong attitude about the place of women in oil geology: "'A woman can do everything in geology that a man can do,' she says confidently.'" (The American Weekly 1945).

King was working out of Jackson, Mississippi. Mississippi was a hotspot for oil discovery in the 1940s (Kennedy 2017). Domestic oil discoveries of oil in places like Mississippi drove the demand of women in the petroleum and gas industry. In Pidgeon's survey of women working in the postwar period, she reported that women in geology, because of the war efforts, would continue to meet the demands of the workforce and achieve continued success: "In this event the women geologists, especially those with graduate training, will continue to enjoy relatively better opportunity than they did in the days before the war" (Pidgeon 1946, pg. 7–7). Though men returning from the war might be a threat to the job force and opportunities for women, women made up a small part of the overall workforce, and predictions indicated that there would not be an oversupply of female geologists. Women were around three percent of the profession according to Mary Elizabeth Pidgeon. A quarter of all the women in geology were working in the petroleum industry in 1946. Women also reported that the industry became more receptive to their place in it after the war,

> A year VJ-day, women seeking jobs in the petroleum industry found that the attitudes of industrial employers had definitely been tempered the war. There was a greater recognition of the ability of women laboratory technologists in research departments, but by and large the opportunities open to women were those which had been open (in few numbers) before the war. Women with backgrounds in geology were encouraged to take clerical or drafting jobs, where they might be especially valuable in assisting staff members in their work. (Pidgeon 1946, pg. 7–10)

However, the publication went on to explain that in 1947 the move from the office to the field really began to occur. King was an example of a transitional figure, but also a woman that represented a new wartime identity for women that was much different than that of Augusta Kemp, discussed earlier in this book.

The profile laid out a list of the tasks that King was performing. The tasks were highly technical, and examined the work from the field: "Working out of Jackson, Miss., she proves that claim by reviewing the logs of wells being drilled, entering the data on maps., sampling rock cuttings and checking data from electric logs which are lowered into the wells to give the engineers an idea of what's going on" (The American Weekly 1945). The emergance of the woman geologist that was both technically savvy and able to provide research perspectives, often with a microscope, but could also get dirty in the field, while maintaining pin-up, or socially desirable, physical features.

She also embodies the characteristics that would emerge in the 1950s as the idea of the "girl next door" or a "one of the guys" dynamic. The notion of the "girl next door" would become explicitly part of the American conversation with the publication of magazines like *Playboy* and the concept of what historian Vicki Howard called "beauty culture" (Kitch 2001; Pitzulo 2008; Pitzulo 2011).

King, in the article, emphasized how enmeshed she was in the culture of the drillers, but how she was considered one of them. The article contained many quotes from King that emphasized her toughness and abilities to exist in the field with the roughest of oilmen: "When in the field, as she is about half the time, she eats with the drillers and makes them feel she is one of the gang" (The American Weekly 1945). The profile then includes a quote from King to emphasize that fact: "'People think that oil crews are a tough, brawling lot of men,' she said recently, 'but I've found them a fine bunch of boys. As soon as they learned that I was working for a living, just like themselves, the stopped thinking it was funny to see a woman in the field'" (The American Weekly 1945).

The profile also empathized that she was "The only female geologist" working in the Carter linked Standard Oil Branch. She was making a lot of money and the profile empathized that she would like advance in this "strenuous career that she has picked out for herself" (The American Weekly 1945). A lot of the profiles of female geologists working in the oil field empathized the self-selection of the career and the initiative to pursue geology education.

Then the article, in keeping with the scholarly idea of Howard with the emphasis on beauty culture, gives her physical measurements: "She's give feet three inches tall, weighs 118 pounds and loves smart clothes, movies and dancing—after work" (The American Weekly 1945). Physical descriptions then continued "A brunette Greer Garson" as she was described by some of the oilmen. But the profile does refer to her in the same sentence, as "...the young Indianapolis-born scientist" (The American Weekly 1945). The oilmen then implied that Garson was lacking, compared to King, because she did not have the scientific abilities of King: "'Everytime I go to town and see Miss Garson in the movies,' one of the boys added, 'I ask myself: What's she got beside those blond locks that our June hasn't got?'" (The American Weekly 1945).

The anonymous oil worker that was interviewed continued with his assessment of King, compared to movie stars. He continued that, "Confidently, this girl out to be in pictures. But we're all hoping she stays right here in the oil business with the rest of us" (The American Weekly 1945). The caption of King, complete with microscope and working in a lab contained the subtitle, "The Pin-Up Girl of the Oil Fields is No Belligerent Feminist, but She Says There's No Reason Why Women Can't Do Everything in the Field of Geology that Men Do" (The American Weekly 1945).

During World War II, pin-up culture was present in the lives of many women. From the noses of B-22 Flying Fortresses to studio advertising to women to provide a pin-up picture of them posed for their romantic partner, women were societally encouraged to participate.[1] The historian Michael West, author of "The Birth of the Pin-Up Girl: An American Social Phenomena, 1940–1946," argued that pin-ups, though most associate them with men's magazines like *Esquire*, were meant to portray everyday women and were popularized in regional and local newspapers, and actually first appeared in *Life Magazine* (West 2020, pg. iii). West contended that the first associations that most people have about a pin-up "...we typically imagine a 'blonde bombshell' actress bursting out of her bullet bra and swiveling her hips to rock 'n' roll, or a long-legged beauty painted by any one of the several dozen pin-ups calendar and advertising artists" (West 2020, pg. iv). But this is not the historical case (West 2020). Originally appearing in 1943, there was a shift, West historicized, "By the end of 1944, and into 1945, pin-up girls were increasingly represented by the girlfriends, wives and even children of homesick soldiers needing a boost in morale" (West 2020, pg. iv).

[1] Examples can be found in newspapers, such as the *Seminole County News* on Thursday June 29, 1944, and in *The Pomona Progress Bulletin* on Thursday August 09, 1945. The Art's Studio of Charm had advertisement for Pin-up photos.

Pin-up type pictures became more common in advertisements, "Pin-up pictures and illustrations were used in advertisement for consumer products or as a promotional material, typically for films and seaside resorts" (West 2020, pg. 12). Though the common assumption with pin-ups was that they purport to display idealized bodies, they did not always adhere to beauty standards: "These images tended to feature the honed bodies of the young and sexy pin-up girls and, in the case of strip tease artists, could also include nudity, but pin-up images were also demure portrait photography of mothers, wives, and children" (West 2020, pg. 12).

Not all pin-ups were focused on large secondary sex characteristics, like breasts: "Pin-up pictures commonly relied on 'cheesecake,' or 'leg art,' to focus the viewer's attention on the model's legs and hips. While the prevailing public image of pin-ups relied on cheesecakes for their sex appeal" (West 2020, pg. 12). However, in the original context, West pointed out, *Life*, was trying to portray aspect of American life, but often included, "….unfailing representation of every American and their penchant for publishing phot-articles of pretty young coeds, Hollywood actresses, models, strip tease artists, sweater girls, as well as letters and photography from readers established the crucial and often under-acknowledged elements upon which pin-up girls rested" (West 2020, pg. 19). Often newspapers took their style and followed *Life*: "*Life* and regional newspapers seemingly acted in dialogue with each other in their willingness to publish material on glamor girls and glamour beauty ideals, often appearing to be in lock step with each other."[2] (West 2020, pg. 19).

The style of King's profile in the *American Weekly*, a special supplement that often appeared with newspapers on Sundays, was an influenced by the pin-up culture of the 1940s in America (The American Weekly 1945). The Pin-up influences could also explain the emphasis on the association between King and the movie star, Greer Garson, pictured in Fig. 4.3.

Garson had a career apotheosis during World War II. She portrayed famous female scientists during World War II, where she was nominated for the Academy award for Best Actress in *Madam Curie*, which came to theaters in 1944 (Troyan 1999). The 1943 trailer billed the film as "The Love Story of The Most Exciting Woman of Her Day! [,]" while also included that it was "The Drama of the Man Who Shared Her Astonishing Adventures."[3] Garson was associated with the oil industry, as she married a wealthy oil businessman in 1949 (Troyan 1999).

The 1943 movie was likely inspiring to many female scientists, or at least on their radar, as the film made 2.575 million dollars in 1943, or $45,413, 662.83 (Troyan 1999; Kentucky New Era 1949.[4] In a movie review, the critic Glenn Hasselrooth discussed the links of Garson's work and popularizing science and as also praised the movie (Hasselrooth 1944). After praising the portrayal of the Curie's relationship, especially its courageous vulnerabilities, he pointed out that

[2] This quote came from page 19, many of the other quotes come from page 11.

[3] Case and font changed for clarity; see the trailer on YouTube.

[4] This was based on several estimates using different websites. Also see the AFI Catalogue entry: https://catalog.afi.com/Catalog/moviedetails/556 and Internet Movie Database entry: https://www.imdb.com/title/tt0036126/releaseinfo/

Fig. 4.3 Greer Garson (Life 1943, pg. 92)

the end of Marie Curie's life, especially the last 28 years does not appear in the film. The critic praised the complex nature of being a woman in science and highlighted Marie Curie's life is important as it conveys those multipart identities: "...the world learned that a life dedicated to science need not ignore the spirt; for Marie Curie never did anything to dim the luminous quality of her being as scientist-citizen of the world, wife, and mother" (Hasselrooth 1944).

Hasselrooth analyzes that most of the important details of the Curies are lanced off in order to promote the dramatic nature of the story and emphasize the romantic nature of the film. He praises the ability of Garson to convey both the scientific talents and the beauty of being a woman,

> Miss Garrison, we know, is one of the most charming actresses alive. Her restrain the laboratory scenes is not that of the selfless scientists at work, but of a beautiful, woman smarting under the stench and fumes of the pitchblende she is cooking, and worrying for fear some of it will splash out on her seductive fascinator. Throughout, she seems addicted to posing dramatically as if Marie Curie know she is making history and will go down as an immortal scientist. This awareness was not part of the real Marie Curie and it is unfortunate that the picture should have been made which even hints that she could have been theatrical. (Hasselrooth 1944)

The critic does admit that the film was a masterpiece. But the review brings a lot of ideas about identity and how outside commentators viewed women doing science.

They should be praised because they sacrifice their beauty to do difficult science, or field work, when they could be doing more glamour jobs. But it implies there is a certain level of glamour now in women doing science. And at the same time, a woman should not look as if she is trying too hard, is too proud, of her scientific work. The duality and pressure were also on women in science to have movie star looks and fashion, while doing the deeply technical jobs associated with science. Hasselrooth's description would summarize Curie, and the way that petroleum geologists like King were narrated in write-ups and pin-culture of female scientists during World War II.

4.3 Ellen Posey Bitgood, and the Female Geologist Asserting Herself in a Man's World

Ellen Posey Bitgood, (pictured in Fig. 4.4) one of the women featured in the newspaper write-up in the previous chapter, attracted acclaim and attention from newspapers for her work in petroleum geology during the war period (1941–1945) and also after the war in the 1950s. The eye catching headline read: "Woman Success at Finding Oil," which ran in the *Oxnard Press-Courier* on March 2, 1953. The paper ran out of California. The article conveyed the excitement of prospecting for oil: "—Ellen Posey Bitgood gets as much kick out of seeing an oil well 'come in' as if she owned it. The gush of the black fuel means that she has scored another success at her unusual job" (Oxnard Press-Courier 1953). By 1953, Bitgood had moved on

Fig. 4.4 Ellen Bitgood working onsite with microscope. (Photo courtesy of Robbie Gries and Bitgood Family)

from her junior role as an unexperienced geologist to an expert. The article proclaimed that "Mrs. Bitgood, who lives here, is one of the nation's leading geologist and specializes in locating likely drilling sites for oil wells" (Oxnard Press-Courier 1953).

On August 14, 1952, *The Daily Ardmoreite* wrote that Bitgood was exceptional, as she was "…the only woman geologist in the world who 'sits on oil wells.'" (The Daily Ardmoreite 1952). *The Daily Ardmoreite* went on to define the phrase to sit on a well. "'Sitting on a well.' In drilling parlance, means watching operations from the well side and analyzing drill cuttings that are brought up with the cutting mud.'" (The Daily Ardmoreite 1952). The newspaper implied the high value of Bitgood's work on the wells: "Analyses better be correct, too, for drilling a well is expensive business." *The Daily Ardmoreite* highlighted that "Mrs. Bitgood, as a consulting geologist, is much in demand" (The Daily Ardmoreite 1952). It also added that she works on the field with her dog and her son. Edward G. Kadane, who was in charge of G. E. Kadane and Sons, an oil company in Wichita, Kansas, praised her work ability "'She's equal of any man I ever saw and better than most[.]'" (The Daily Ardmoreite 1952).

The *Oxnard Press-Courier* article, in 1953, continued to promote the novelty of a woman working in a mostly male profession (Oxnard Press-Courier 1953). The article praised her efforts and her training, but also highlighted the long history of men in the petroleum geology profession, implying that it was monopolized by men because it took both field and academic knowledge,

> She's one of the few women doing the job that men long have monopolized, and with good reason. Searching for oil is a demanding occupation.—for it takes years of scientific training, lots of scrambling over rough country, and an encyclopedic knowledge of rocks and fossils. (Oxnard Press-Courier 1953)

The article then gives her biography of a family that many of the male members worked in the oil business, and encouraged her entrance into the petroleum geology profession. Her father and uncle were involved in the oil business. Her father was a "cable tool driller," which was a job that involved drilling into an oil well with cables to destroy any rock in the way of removing the oil.[5] Her uncle was also very involved in getting Posey to focus on a career in petroleum geology. Bitgood, then Ellen Posey, stated that she was planning to major in journalism at the time, but was talked out of pursuing that profession.

The named uncle, J. P. Wolfe, promoted geology to Posey, because of the potential jobs for women, as the field was opening up to women. Wolfe's encouragement was recorded in the article through Posey, "But her uncle talked to several geologists—men—who assured him there was a great future for women in the field" (Oxnard Press-Courier 1953). She then studied geology at the University of

[5] See the glossary: https://glossary.slb.com/en/terms/c/cable-tool_drilling#:~:text=A%20method%20of%20drilling%20whereby,the%20side%20of%20the%20tool

Oklahoma and graduated in 1930. Her first job was with Cities Service, late renamed CITGO.[6]

A former executive that Posey worked for praised her capabilities and successes in the field, "'She was a sharp one,' an executive recalled." The executive remarked that she was treated differently as a woman, but she persevered through her fellow male employee's treatment,

> It was a new experience for the men, having a woman out on a drilling rig. And some of them had to have their practice jokes. They'd doctor her drill samples—once someone went down to the river and got a little sand and slipped it into a sample bag just to see what would happen. (Oxnard Press-Courier 1953)

The article then abruptly changed to praised her domestic skills as a woman and homemaker. The subtitle in the article included "Good Cook, Too." The subtitled section continued with Posey's response of sarcasm to the male sabotage of her work, with the same executive interviewed discussing Posey's response,

> In a few minutes we heard her whistle—she could whistle so you could hear it a half mile—and saw her wave her arms. She looked up from her microscope and said, 'Well boys, I don't know what you can do about it, but the Arkansas river bed has just moved into the bottom of that hole. (Oxnard Press-Courier 1953)

The executive recounted the energy that Posey returned to her male co-workers. The executive included that, "'No matter what trick the men tried—they never tripped her,' the official added. 'Pretty soon they gave up trying.'" (Oxnard Press-Courier 1953).

Then the article included information about Posey, now Bitgood's personal life. She married in 1936 and quite her full-time job at Cities Service. She works now as a consultant and has a son who was 11 years old at the time of publishing the article. The article ends with additional information about her domestic life. The unnamed writer of the article recorded that, "Mrs. Bitgood has a reputation for being a good cook, and the word is that the men on the drill force liker her around for that reason" (Oxnard Press-Courier 1953). Bitgood's value to the team, at the end of the article, was linked to her domestic skills.

The expectations on women, both during the war, and after the war, were multi-faceted and no doubt exhausting. Male expectations on women were that they had to adapt to technical mastery of wartime needs, but also had to maintain domestic excellence in cooking, cleaning, and homemaking. The sociologist Maria Cristina Santana wrote about the duality of the experience of women during World War II, and the public and personal expectations put upon them (Santana 2016). Women in many countries, including the United States, Canada, and the United Kingdom, entered into the workforce to replace male labor. Many women served in munition factories, oil fields, medical care, war service on the frontlines, scientific, research, and in other ways. Many women served in these professional and industrial capacities out of economic and nationalist interests. Santana contextualized that "Women's

[6] See the website for the history of Cities: https://aoghs.org/old-oil-stocks/cities-service-company/

work in the war industry was seen as an extension of their domestic duties" (Santana 2016, pg 3). At the same time, she pointed out, many women married, often earlier than expected, out of "patriotism." Jobs were plentiful for women during the war, and she recorded the experiences of women who lived during this historical period, as they stated that no one had to encourage women to take war related jobs, as they said they took these jobs out of the spirit of serving their country.

But the ideal image of womanhood still haunted and defined expectations of women, even while they were taking on new roles in society. Feminism was still being defined by past expectations: "During the war, female masculinity provided for an expansion of people's ideas of what women should be and how they might look and behave 'leaving a canon of images to inform future versions of feminism…" (Santana 2016, pg 3). And though images like Rosie the Riveter would define this new woman, those working women were given words of caution about their changing role in society. Santana found that, "One article in the Navy shipyard newsletter counseled women to 'be feminine and ladylike even though you are filling a man's shoes'[.]" (Santana 2016, pg 3). The messages were often contradictory, "Women were receiving confusing messages, such as their instruction in charm courses at Boing [sic Boeing?] airplane plants. This temporary employment allowed women to dress differently, behave in different ways, and allow for different career choices" (Santana 2016, pg 3).

The challenge of taking on literal new clothes, and expectations in society, while maintaining the female ideal must have been changing to women who took these new jobs. Ellen Posey Bitgood continued to have a type of geology celebrity, and the media both praised her mastery of her job as a geologist, while at the same time praising her femininity. These trends continue in the many articles about Bitgood, as well as similar articles to the *Oxnard Press-Courier* that run in other papers.[7]

Bitgood's story was also picked up in Spanish language papers, such as in the paper *La Opinion: Diario Popular Independiente*, which was based in Los Angeles. The article "Una Dama Que Ha Hecho Surgir Mas Petroleo Que Muchos Hombres," which translates to "A Woman Who Has Drilled More Oil Than Many Men," was published on July 17, 1953, profiled Bitgood. *La* Opinion in its section "Femeninas" ran a similar story as the *Oxnard Press-Courier*, but it contains more information and a different ending. The article also emphasized more of the opinions of her male family members (La Opinion 1953). The article ends differently than that of the *Oxnard Press-Courier* piece, as it closes with, "Ellen Bigtood haaberierto una nueva frontera parala expansion de las actividades femeninas en ua aspera campo, sin perder nada del ecnato de su sexo" (La Opinion 1953). This sentence translated to, "Ellen Bigtood has opened a new frontier for women in a rough, male-dominated profession without giving up any of her feminine charm." But the article contained the line about Bitgood's domestic prowess: "Mas admirable aun es que Mrs. Bitgood, encuentra timepo todavia para atender su casa y ser model de esposa y

[7] Other papers that run a similar version of the article include, but are not limited to *The Yuma Daily Sun* on July 20, 1953, *Lubbock Evening Journal* on March 11, 1953, *Lubbock Avalance-Journal*, and other newspapers run similar stories on Bitgood.

madre" (La Opinion 1953). And the article ended the same way as the other newspaper articles with the praise of her technical, as well as feminine virtues: "Even more admirable is how Mrs. Bitgood is able to find the time to be a model wife and mother at home."[8] (La Opinion 1953) In 1953, the article ran from California, Delaware, to Texas to Florida. Each article had a promotion about the novelty of women working in the petroleum industry and in the field. Some titles included "Woman Masters Man-Size Job: Oil Wells Are Her Business" (The Sunday Star 1953).

The Panama City News-Herald ran a similar version of those that ran in Texas and California. The article emphasized the comfortableness that Bitgood had with dealing with rough oil workers. Her office co-worker, John A Kay, recalled that "…he long since became accustomed to seeing a big, two-fisted derrick worker come into the office looking for 'Miss Ellen.'" (Panama City-News Herald 1953). Kay continued that "'After a little foot—shuffling,' Kay said, 'he will usually say something like this, 'Miss Ellen, we want to know when you're going to be on another well for us. We ain't had a good cup of coffee on the rig since last time.'" (Panama City-News Herald 1953).

The article continued the trend of writing about women with scientific and technical expertise. The skills were praised along with the additional and dual praise, as well as implied expectation of proficiency of domestic skills, such as making coffee or prowess in cooking. Women like Bitgood had to excel at their field, while making male co-workers and clients feel welcomed and cared for, and the duel expectations followed each other in publications.

4.4 Two Girls as Oil Detectives Dorothy and Doris

There were many other women like Ellen Posey Bitgood, who were praised for their technical experience and mastery, but were also defined by their socialized and gendered characteristics. In 1941, the *Saint Petersburg Times*, published out of Florida, ran an article promoting female success in geology: "Two Girls Achieve Success As Geologists in Texas" (Rives 1941). Like the articles promoting Bitgood, these articles ran in various forms across many different newspapers. Some newspapers like the Decatur Herald and Review covered their work with eye-catching headlines meant to draw the reader's attention to their attractiveness, then their scientific achievements, see Fig. 4.5

The article published in the *Saint Petersburg Times* profiled the work of Dorothy A. Jung and Doris S. Malkin. The article was originally written in Houston by William T. Rives. Rives described Jung and Malkin first by their physical features: "The tall, shapely blond and the tiny brunet, strikingly dressed to accentuate their

[8] Spanish translation work was kindly performed by Dr. Mark Groundling of the Spanish department at Tennessee Technological University.

GIRLS BECOME OIL EXPERTS

Dorothy A. Jung, 24 (right), and Doris S. Malin, 27 (left) study a geology problem at Houston, Texas, following completion of a technical research report on marine sedimentation and oil accumulation on the Gulf coast. They left Brooklyn, N. Y. two years ago and have invaded a scientific field largely filled by men. (ASSOCIATED PRESS PHOTO)

Brooklyn Beauties Leave Flatbush for Oil Field

Fig. 4.5 Jung and Malkin pictured working "a geology problem." The headline appeared slightly below the picture (The Decatur Herald 1941)

beauty, might pass as photographers' models." The article then adds the narrative twist. The article continued that, "But they [Jung and Malin] are't. They are oil detectives—geologists who have just completed a technical research report on marine sedimentation and oil accumulation on the gulf coast" (Rives 1941). The article, then takes another approach, profiling the women as finding nothing unusual about their choice of profession.

Rives described the women as "invading" the field of petroleum geology that was male dominated. To Malkin and Jung, they were comfortable in their chosen profession, "Blond Dorothy A. Jung and diminutive Doris S. Malkin find nothing unusual in the fact they have invaded successfully a field generally monopolized by men" (Rives 1941). The women were quoted on their position in their profession: "'Why shouldn't we be geologists? They chorused. 'Women geologists sometimes are better than men.'" (Rives 1941). They changed the status quo in geology, but the war was seeming to fan those existing sentiments and make them palatable to the public.

But female geologists had to maintain their gendered roles in society as well (McEuen 2010).

The article pointed out there many professional memberships in oil and geology–related societies, but noted that there was a small amount of female membership, such as in the Society of Economic Paleontologists and Mineralists. The article noted that there was a rising number of women in these societies, including 31 female members. The two women did acknowledge their challenges in the field, they specifically noted that they received a lot of bad treatment in the field because some men did not agree with their role in the workforce. The article relayed their experiences: "And Dorothy and Doris admit its pretty tough on the women at times, being a geologist. Men who feel the female should stay in the kitchen sometimes resent a woman intruding herself upon a he-man profession" (Rives 1941). They were warned when they arrived in Teas from Brooklyn that they, as women, would not find jobs.

However, their persistence and attitude defied the status quo. The article cheered that, "But Texas just weren't familiar with the Brooklyn grand of persistency. It took Doris a year to land her job, and Dorothy's money virtually ran out before she caught on with a production company." The women found jobs despite difficulty challenges, as Dorothy recalled that, "'I had no money to return to school and study for my doctorate,' Dorothy says, 'so I made up my mind to stay here until somebody put me to 'work.'" (Rives 1941). Dorothy had to work two years before she was able to work in the oil field. Dorothy mentioned that she was not able to "sit on a well'" for a long time, a task that would require her to work for eight to ten hours in the oil field, examining the well for its long-term profitability and oil production. She summed up her experiences as that "'they felt a woman just didn't belong out there.'" (Rives 1941).

Dorothy was 24 at the time of the writing of the article, while Doris was 27. Both women found an interest in geology in college. Doris mentioned specifically her interest was peaked during her childhood in girl scouts. She remembered fondly that during her childhood "she had 'studied rock while I was a Girl Scout. In fact, I have my rock finder's merit badge.'" (Rives 1941).

Doris was working on a doctorate in geology, and it was noted that Dorothy playfully called her "Doc" but only among themselves. The name was a point of pride for the two women and indicative of their positive relationship. The article reported that, "…Dorothy called her 'Doc'—but only in the privacy of their conversations. 'When we're out on dates, she insists I call her by her right name,' laughs Dorothy" (Rives 1941). But the conversation quickly returned to the sexism inherent in geology at the time and in the oil offices.

The article also mentioned that Dorothy had a nickname as well. The nickname was vulgar and offensive, especially given the time period. The article noted that "But Dorothy, too, has a nickname the office force tagged on her as soon as they found out she has from Brooklyn" (Rives 1941). The nickname was "Flatbush Floogie," and upon revealing the name Dorothy's reaction was recorded that "…she sighs sadly" (Rives 1941). The Flatbush referred to the neighborhood in central Brooklyn, and term "Floogie" was a shortened version of "floozy" which could mean anything from a prostitute to a woman of loose morals. The term was recorded in several songs of the time and was a term that was extremely unkind, and was term that was incredibly offensive to a co-worker. The nickname continued to show the duality of treatment of women in geology, especially in the oil industry, where men could cut down at their authority at any time.

Dorothy Jung Echols did eventually become a full professor, after working out west in the oil industry, from 1938 to 1951, at Washington University in St. Louis. The memorial went on to discuss the friendship of Doris and Dorothy. Doris went on to finish her Ph.D., who was a fellow graduate student, who went on to become the first woman of the Geological Society of America. Doris recorded a memorial to her friend Dorothy, which was recorded in the remembrance,

> She brings a spirit of learning, inquiry, knowledge, wisdom, and adventure to the classroom and the laboratory, whether it be in the introductory level class for non-majors or at the Ph.D. research level. Her students want to learn, they do learn, and they become successful productive scientists, or lovers of the earth. They leave her rouses with a built-in lifelong scientific attitude, and a loyalty to the earth sciences and to 'Mrs. E.' Her personal, caring approach to teaching fosters these attitudes. (Price 1997)

The women go on a publishing streak during the war as well. From 1941 to 1948, the two women publish three articles together. These papers included the mentioned publication in the article from 1941 about Marine sedimentation that was published in the *American Association of Petroleum Geologists Bulletin*, and two other post-war pieces that appears in the same publication in 1948 (Price 1997).

Both of these women were highly competent geologists, who were working for the war effort. But they were marketed by the Associated Press for their physical appearance and femininity first, and then their scientific achievements, education, and persistence to participate in male-gendered work like petroleum geology followed in the stories (Rives 1941; Decatur Herald and Review 1941). However, women explained using non-threatening that they were not treated equally as Dorothy was called a "floogie" and "sigh[ed] sadly" (Decatur Herald and Review 1941). The ability to participate in science was taken advantage of by women like Jung and Malin, but they had to tread through that world through the view of womanhood and femininity, even during World War II (McEuen 2010; Jack 2009). Articles promoted the glamour of doing things like "invad[ing] Men's Field" but were not treated seriously by some of their colleagues, and were put down with sexist language.

4.5 Higher Education Meets the Wartime Demands of Petroleum Geology with Female Geologist

The line between women's work and men's work unravel through the years of World War II. The War years created a shortage of labor and removed the surplus of works that had occurred in the 1930s (Education for victory 1942). Jobs in the oil industry needed educated workers, and streamlined educational programs and opportunities were opened to women, especially in 1943 to 1945. In the publication *Education for Victory: The Official Biweekly of the United States Office of Education Federal Security Agency*, an article appeared on March 3, 1942, which promoted the usefulness of training women for the petroleum industry to help the wartime effort. The article, "Training Women for Work with Oil Firms," promoted that at the University of Michigan that the school would be offering a 12-month accelerated course for petroleum geology (Education for victory 1942). This program, designed for women, would make them useful for "…the discovery programs of American oil companies" (Education for victory 1942). The chairman of the geology department, Kenneth K Landes, wrote that the new program would be both academically and practically based for women who wanted to gain access to careers in oil. Usually the academic program would take students four and a half years to complete, but this accelerated track would only require one year's study (Education for victory 1942).

The program would condense all the sophomore, junior, and senior classes into the programs' one-year plan of study. Students had to enter with some geology credit, such as classwork prior physical and historical geology. At the end of the war, *Education for Victory* promised, the students could return to the University of Michigan for additional classes and receive their degree. The article stated that 20 to 30 women were planning to take the course and oil companies promised to hire many of these women, as well as men who were not eligible for military service, could also complete the program and pursue employment. The program included a field work component that could take places over the summer session in 1943 at Camp Davis, which was located in the city of Jackson Hole, Wyoming. The approach of the accelerated program, according to the chairman, was not as much focused on education as practical training, as he pointed out that in petroleum geology that was a "serious personal shortage" (Education for victory 1942).

The Michigan Daily, the student newspaper of the University of Michigan, ran an article in November of 1943 that mentioned that female geology students were interviewed by oil companies. The article "Oil Firm Interviews 'M' Co-ed Geologists," which was likely a reference to mineral geologist, explained that some of the first graduates from the accelerated geology course were being interviewed by oil corporations (The Michigan Daily 1943a). The article promoted the success of the program, "Nine women, who will be graduated at the end of the present semester in the concentrated program in petroleum geology, were being interviewed by a representative of an oil company this week" (The Michigan Daily 1943a). The chairman of the program, Kenneth K Landes, was interviewed for the article and emphasized the importance of women serving in the petroleum industry and are in demand in the industry: "'Other oil companies are already showing interest in the girl geologists as future employees and are sending interviewers here…'" (The Michigan Daily 1943a). The class of women, who started the program in February 1942, were approaching graduation and likely had jobs waiting for them in the oil industry.

The Michigan Daily continued to report and promote the program. On January 3, 1943, the article "Petroleum Geology Course Offered Next Term To Equip Women to Handle Essential War Work," which ran among other articles that discussed women and their roles for employment in industries that supported the war, discussed the geology program in depth (The Michigan Daily 1943b). Chairman Landes continued to promote the direct access that the program provided to women in joining the petroleum industry, "'This special training program, set up at the request of a number of leading petroleum concerns, will provide in the one year period as much academic work in geology as students preparing for the profession of petroleum geologist ordinary obtain in four and a half years'" (The Michigan Daily 1943b). Landes continued to promote the program as filling industry shortages, but this time giving the a similar quotation but emphasizing the importance of the war, "'The program is not intended to provide an 'education'...but rather to give essential technical training in a field in which a serious personnel shortage exists because of the war.'" (The Michigan Daily 1943b).

The need for more people in petroleum discovery was because of the heavy demand on the oil stores on the United States because of World War II. New fields of oil had to be discovered in order to sustain the long-term needs of the United States. *The Daily Michigan* explained the oil emergency to the readers, though there was a typographical error, and the end of the quote was not marked: "'The petroleum 'discovery curve' in this country is in its fourth year of sharp decline.[‘] This means that the United States id drawing heavily on reserves and that many new fields will have to be discovered in the next few years if we are to continue to be self-sufficient, or practically so, in oil production" (The Michigan Daily 1943b). The chairman of the geology department continued to remind readers that oil exploration was one of the main functions of petroleum geologists. They were in demand because of the stress on existing employees because of war service and increased demand for oil to fuel America's military.

The article went on to explain that the oil companies were reaching out to Michigan in order to get training programs for women in order to fill their personnel needs. The letters requesting assistance were sent with the highest possible speed, "Within the last two months letters or phone calls have been received from seven major oil and gas companies requesting women trained in geology for positions on their staffs. Air mail letters were sent to the chief geologists of thirty-five oil companies requesting their reactions to the plan. Replies from half of these men have already been received an included promised job for twenty women" (The Michigan Daily 1943b).

Penn State, for example, wrote proudly in 1944 that their "Extension services" had seen high enrollments of both men and women serving the war effort (The Pittsburg Press 1944). They had about 4000 students a year training for careers offered through their Mineral Industrial School. These correspondence courses were in high demand for both male and female students. The United States Armed Forces Institute allowed access to training materials from ceramics, mining, geology, gas engineering, and petroleum refining to serving women and men, as well as civilians wanting to serve the war efforts (The Pittsburg Press 1944).

The War brought women into the forefront of geology, as they were put to work for the war effort. And society, and institutes of higher education, opened up services to help female students. Women working in geology, especially those involved with the petroleum industry and the war effort. Women were portrayed as heroic, as well as feminine, but industry sometimes embraced women defying societal identities of "womaness" that were placed upon them.

4.6 "These Women Aren't Sissies, Either!"

The newly educated woman was portrayed in both advertising and recruit posters, which emphasized both their war service, the temporary nature of their jobs, but also the idea that they had not abandoned their home and woman-focused responsibilities. Maureen Honey framed the ideas associated with wartime "Womanpower" in women's relationship to both wartime service and home front responsibilities, "We find, therefore, many advertisements which implied that female defense workers had not abandoned home responsibilities, nor their aspirations to be full-time homemakers. Women in factory overalls were frequently portrayed in the home doing laundry or cooking meals" (Honey 1980, pg. 50). Working women were often framed as keeping all industries running until men could return home from the war, and keeping the home front steady (Honey 1980, pg 50). One industrial worker was portrayed in a 1944 poster in the *Saturday Evening Post* as the newest Molly Pitcher. Pitcher was involved in the American Revolution supporting men on the battlefield. Many advertisements emphasized woman serving in uniforms, but also the duality that they also wear kitchen aprons. Honey contextualized this duality: "A technique which minimized the threat these egalitarian images posed to traditional feminine norms was the use of historical models which placed war work in a context of female loyalty and support" (Honey 1980, pg. 52). Therefore, readers came away with the concept that, "As portrayed by propagandists, then, the motivation for women in war jobs was not money or status or job security, but patriotism and love" (Honey 1980, pg. 52).

Standard Oil put out a marketing effort to praise their female workers, alongside their male workers during the War and Postwar years. Since around 1927, companies like Standard Oil talked about the pride that they had in their employees, both men and women. In advertising the Sunday classical music radio program, the company praised its female workers. The advertisement contained both notions of the future of women's work in the oil industry, but also justified their continued success through the utilization of history.

The advertisement was meant to drown out any opposition to women in United States oil production. The ad promoted the idea that women were as a good as any male engineer, "*No sissy* business is the job of riding scientific herd on an oil rig that's drilling under a desert sun. That's the job of a petroleum engineer. Yet, one of our petroleum engineers is a woman" (Spokane Daily Chronicle 1945). The ad continued to promote its female engineers: "*No sissy* business is the job of teachings tricks to those elephant-big mechanical gadgets in an oil refinery. That's the job of a mechanical engineer. Among our mechanical engineers is a woman" (Spokane

Daily Chronicle 1945). The ad featured a working female geologist confidently working in the oil field, as pictured in Fig. 4.6. Then the ad turns to historical examples to prove the overall toughness of women, "*Well*, come to think of it, women never were sissies, anyhow. Look at the Revolution's Molly Pitcher. Look at the Civil War's Clara Barton. Look at World War I's Edith Cavell. Look at the thousands of woman who, in this war, are doing men's jobs."[9] (Spokane Daily Chronicle 1945)

The ad then shifts to make the link between the war service of historical women and the women currently working in the oil business, and standard oil as well: "*But look* as you may, you'll find few doing more *important* work than the woman engineers, chemists, architects, and technical experts now holding down key posts in our organization. With their special knowledge and training, they're helping keep the oil flowing to the armed forces—and to you" (Spokane Daily Chronicle 1945). The ad then embraces the feminine beauty standards and lays out the contrasts in character of these female workers, very much like many of the other female geologists and oil workers featured in this chapter, "*This may not* surprise you as a much as it does us. That's because you get small change to look beyond the service station and see how largely the oil business consists of overalls and efforts. Oil gets under your nails. Sweat gets in your eyes" (Spokane Daily Chronicle 1945).

The ad continued to emphasize the importance of women keeping the War oil industry going: "*Yet*, as many pivotal points, that un-sissy process whereby a barrel of 'crude' under the ground in California gets to the Philippines as 100-octane gasoline is kept going by women" (Spokane Daily Chronicle 1945). The ad ends by promoting their introductions on the radio show, sponsored by Standard Oil, entitled "'The Standard Hour.'" The end of the ad equates female oil works with their importance to the war effort and their toughness: "*They're* making careers in an industry

Fig. 4.6 Image attached to the Standard Oil Ad that ran in the *Spokane Daily Chronicle* March 20, 1945, proclaiming that "These women aren't sissies, either[]"

[9] I reached out to the Stanford Archive of Recorded Sound that held some of the programs but could not find this specific program. Unfortunately, these radio programs were only recorded by listeners who felt inclined to record it, and there are no recordings that I have been able to find of this program.

almost as masculine as the Army. They're doing a 'must' job. We thought you would like to know about them" (Spokane Daily Chronicle 1945).

4.7 *The Petroleum Engineer*: Women, Society, and World War II

Not all in petroleum geology were comfortable with women entering into the workforce and being present in oilfields, especially prior to World War II. *The Petroleum Engineer* was a trade publication that was in publication since the early twentieth century. Prior to the war, the publication mostly discussed women in terms of jobs, such as the 1938 edition that contained the section, "Laugh with Barney," which included jokes like "Men are particular, just as women who have long suspected. For instance, a fellow who hadn't kissed his wife in five years, shot a fellow who did" (Horrigan 1938; Giebelhaus 1996; Croneis 1933). However, after the war the publication ran a section entitled "Women at Work" which promoted the work of women in the oil business. The magazine was in a juncture during the late phases of World War II, as men in the industry were conceptualizing the new reality that they found themselves in with women in the work force. Perhaps the images utilized in the article were meant to soften blows to masculinity, as Maureen Honey argued in the study of World War II posters and woman in the work force (Honey 1980).

In 1944, *The Petroleum Engineer* ran an article "From Dimity to Denim" which took efforts to discuss the new female worker in the oil industry. The name of the article is an allusion to the more Victorian dresses of many layers and harnesses to the denim plants of the wartime female worker (The Petroleum Engineer 1944a).[10] The article discusses the new life and society that female works now find themselves in:

> Milady of yesteryear scarcely would recognize the fashion parade of today. A new wartime has swept the nation and stolen the show. Work slacks, anklet sox, tight headwraps, safety shoes, and heavy gloves have come into style as fashionable American women trade dimity for denim and go to work win this country's war plants. (The Petroleum Engineer 1944a, pg. 134)

The article continued to historicize the role of women in the car through their clothing and dress. The article also tried to create a narrative of women as being delicate and non-masculine prior to the start of the war through the contrast of clothing: "Such a mode wasn't dreamed of in milady's era of silk and satin. But World War II caused many changes—and America met each in stride. The switch in industry where denim-clad women take over jobs formerly done excessively by males is one of the great social adjustments of the war" (The Petroleum Engineer 1944a, pg. 134).

---

[10] *Merriam Webster* defines dimity as "a sheer usually corded cotton fabric of plain weave in checks or stripes." In this article, it is meant to refer to old fashion, or old types of traditional clothing.

The article then promoted the transitions of women, even against the skepticism of industrial leaders. The women had left the offices for fieldwork: "At first, some businessmen were dubious. They weren't sure woman could master the muscular tasks of big industry. True, they conceded, women had done well in offices, stores, and professional fields such as medicine and law, but for the so-called weaker sex actually to shoulder the nation's production wheels—well, maybe" (The Petroleum Engineer 1944a, pg. 134). Women stepped up to fill the "manpower shortage" and got the opportunity to transition into new career roles and occupations in the all industry, as well as the "…rugged oil industry." The article then turns to a promotion to the work performed at Continental Oil Company, which was located in Oklahoma. Prior to the war there were 28 women working the company in testing and the laboratory. But as of September of 1944, the company has seen their numbers include more and more women. Using an allusion to the fierce Amazonian warriors of classical mythology, the article included that at Continental Oil there were a lot more female employees at work for the war effort: "Today, the same visitor would be astounded if he were to stand at the refinery entrance and watch Continental's industrial Amazon army of 300 going on and off work shifts" (The Petroleum Engineer 1944a, pg. 134). The company at present had 1031 female employees and they performed their "…tough jobs like real all-Americans." Women were working anywhere from the office to the field.

The article then shifted to promote the work of some of the female oil workers, and their consciousness of the efforts to help their families at home and their husbands involved in the armed forces. The article also promoted their roles as mothers, as well as employees. The article framed a transition from generalities about historical change to the experiences of actual female employees in the petroleum industry. The article rhetorically asked: "Who are these wartime Eve Curies? From where did they come from? And why?" (The Petroleum Engineer 1944a, pg. 134).

The article then proceeded to answer its question with empathizing the female virtues of the employees:

> They are mothers, wives, sisters, and sweethearts. Nearly all have men serving in the armed forces. Before Pearl Harbor these women war workers were engaged in varied activities. Many were housewives. Others were offices workers, clerks in stores, school teachers, waitresses, high school and college students. (The Petroleum Engineer 1944a, pg. 134)

From female truck drivers to stencilers, each woman was projected, through the article, to explain her place in "a masculine industry," is that they emphatically wanted to win the war: "We want to help win this war." The article argued that despite the money, women were motivated to work 8 hours a day in difficult jobs, such as Mildred Walker working a crane. Her bosses described her job performance as "Swell[.]" But the importance of women returning to the home and leaving the oil industry was also present in the article, as Mildred Walker summarized her plans after the war: "After the war, the woman crane operator plans to resume the household tasks she has left temporarily. But first comes victory" (The Petroleum Engineer 1944a, pg. 134).

The article also emphasized that most of the women that worked the pipeline were working to get their romantic partners home. The roles of women in industry were portrayed as making the return home for male veterans more comfortable. An example can be found in the portrayal of Mrs. Golda Mooney, "Mrs. Mooney, mother of two young children, is convinced that her job of filling barrels of oils and checking petroleum shipments at Continental's C and P plan will help hasten her husband's home coming. And, when the sergeant retunes, he is going to find that his wife has been postwar planning. With the aid of money earned by filling oil drums, Mrs. Mooney has completely paid for a home that she started to buy about a year ago" (The Petroleum Engineer 1944a, pg. 134).

> Though the oil industry started to add a small amount of women, for example the 15 female employees at Continental's C and P plant in 1943, the major influx of female workers occurred in 1944: "Subsequently, home makers replaced men on such jobs as yardwork and maintenance, canning [,] machine operators, barrel cleaners, warehouse helpers, and tool room attendance—to mention only a few of the arduous jobs. (The Petroleum Engineer 1944a, pg. 134)

As many other women were profiled in the article about their jobs tasks in oil, there was a notion of temporality with every interview, ending with: "'When peace time comes,'"…."'I'm going to take over my old job taking care of the children.'" (The Petroleum Engineer 1944a, pg. 134). However, many women continued to move into the geology and petroleum–related industries and some companies were committed to keeping them employed.

In that same issue there were mentions of "Postwar Planning," by oil executives like Harry T. Klein, who was president of the Texas Company, that made promises to include women in employment. The mention made promises that "'The Texas Company intends to go far beyond its legal obligations in placing men and women who return to company service form military leaves of absence" (The Petroleum Engineer 1944b, pg. 130). And the mention makes it known how proud the Texas Company was of both its male and female veterans: "'More than 5,000 men and woman employees of the Texas Company and domestic services are serving with the armed forces" (The Petroleum Engineer 1944b, pg. 130).

4.8 Black Women and the Lack of Geologic Education

Historians can only see snapshots of minority women and men contributing to the geo-sciences prior to the 1960s and the Civil Rights Movement in America. Many current historical studies found that African Americans, and other minority groups missed out on the educational and economic advantages that came to white Americans after World War II (Temin 2022; Altschuler and Blumin 2009; Steiner-Fleury 2012). For instance, black Americans were excluded from the G. I. Bill that allowed for many white men to enter college and advance into technical careers Temin 2022; Altschuler and Blumin 2009; Steiner-Fleury 2012). Also the United

States Government had been segregated since 1913, and the army was segregated until 1948 (MacGregor 1981).

In the book *The Employment of Women in the Early Post War Period*, discussed in this chapter, is an example an occasional glimpse into black women and their work in World War II (Pidgeon 1946, pg. 1-58–1-62). The study commented on African-American women, as well as African-American men, working in science and the petroleum industry, "Although industry and Government have been difficult fields for Negro scientists to enter, Negro women are known to be employed as seed analysts, technical, or chemistry in at least two regional laboratories of the Federal Government and in one laboratory in Washington, D. C" (Pidgeon 1946, pg. 1–60). The publication made specific interest in the oil industry, "A Midwest university in 1946 placed with an oil company a young Negro woman chemist who was married and whose husband was studying medicine" (Pidgeon 1946, pg. 1–60). The publication also referred to efforts by African-Americans and wartime education, "An unknown number of Negro women took chemistry courses under the Engineering, Science, and Management War Training Program and worked in war industries. One who took a 10-week analytical chemistry course at a woman's college under this program was reported to be a 'whizz on carbons.'" (Pidgeon 1946, pg. 1–60). No mention of female geologists is included in other publications like *Negro Women War Workers* that was also published by the Women's Bureau in the United States Department of Labor (1945).

The publication went on to point out that training in geology was often not available at universities serving African-American students (Warren 2000). Between 1876 and1943, only one African-American woman received a Ph.D. in geological sciences at that time and it was likely Marguerite Williams.[11] The publication ran the picture of the unnamed scholar (likely Williams), pictured in Fig. 4.7.

At the tail end of World War II in 1946, Thomas W. Turner, examined science education at predominately black universities and found that the science offerings were insufficient to student needs and wartime demands (Turner 1946). In *The Journal of Negro Education*, he framed the problem that the colleges were not preparing students to make the transition into the scientific and industrial jobs of the United States. Turner noted that most educational offerings at African-American serving universities focused on math, biology, chemistry, and physics, and he directly criticized geology education, "Geology and astronomy are generally omitted. The teaching carries very little or no experimental opportunity beyond the exercise of the laboratory period. The student has little change to learn was productive work in the larger fields of his school activities means" (Turner 1946, pg. 38).

Turner made the point that the stakes of omitting sciences like geology were quite high and hurt the economic opportunities for black students, "The extraordinary applications of chemistry, physics, mathematics and the technological fields based upon them are well known from the dominant-r[o]les they are playing in the

[11] For a good but brief biographical entry for Williams, see *Black Women Scientists in the United States* on pg. 267.

Fig. 4.7 Picture of Williams (unnamed) in *The Employment of Women* (pg. 1–59)

present war" (Turner 1946, pg. 41). And Turner frames the importance of not having proper education in geology or other physical sciences as an opportunity lost for a African-American university graduate, "In neglecting the physical science the Negro colleges are not only ignoring the trends pointing to the largest field of job activity for the future, but they are neglecting also the present day observations as to what science equipment is most successful for earning a living at present and in succeeding generations" (Turner 1946, pg. 42).

4.9 Conclusion

Women working in the petroleum industry, learning about oil and geology, or desiring to serve the war effort entered into a new and exciting frontier, but a frontier that was sold to the public with a duel message of technical mastery and femininity. Women working in geology and petroleum industry were praised for their attitude, patriotism, technical abilities, and physical appearances. Praises of physical appearance and plans for a return to domestic work appeared in print ads and profiles in order to not disrupt the status quo domestic gender roles in America. However, it seems that oil companies made the transition to embrace a lot of their female geologists, especially as petroleum geology became more technical. However, white women working in petroleum and geology seemed to benefit from the spaces that the war opened, but black women were not able to capitalize on the war.

References

(1938) Women enter geology: more and more take up work, some being employed in oil fields. The Evening Republican. 16 December
(1941) Girls become oil experts: Brooklyn beauties leave Flatbush for oil field. Decatur Herald and Review. 23 April
(1942) Training women for work with oil firms. Education for victory: official biweekly of the United States office of education federal security agency 1 (March): 22
(1943) Life calls on Greer Garson in Hollywood. Life 12 April, pp. 90–93
(1943a) Oil firm interviews 'M' co-ed geologists. The Michigan Daily 54(10). 12 November
(1943b) Petroleum geology course offered next term to equip women to handle essential war work. The Michigan Daily 53(66). 3 January
(1944) A college goes to war: Penn state brings learning to thousands via extension. The Pittsburg Press. 19 May
(1944a) From Dimity to Denim. The Petroleum Engineer 15(13):134–136
(1944b) Postwar planning. The Petroleum Engineer 15(13):130
(1945) Negro women war workers. Women's Bureau, Washington, DC
(1945) These women aren't sissies, either! Spokane Daily Chronicle. 20 March
(1945) Mixing glamour and oil. The American Weekly. 23 September
(1952) The Daily Ardmoreite. Here and there in southern Oklahoma. 14 August
(1953) Woman success at finding oil. Oxnard Press-Courier. 2 March
(1953) Femeninas: una dama que ha hecho surgir mas petroleo que muchos homres. La Opinion. 17 July
(1953) Woman masters man-size job: oil wells are her businesses. The Sunday Star. 15 March
(1953) Women's work. Panama City-News Herald. 2 March
Altschuler G, Blumin SM (2009) The GI bill: the new deal for veterans. Oxford University Press, New York
Croneis C (1933) Early history of petroleum in North America. Sci Mon 37(2):124–133
Giebelhaus AW (1996) The emergence of the discipline of petreolum engineering: an international comparison. Icon 2:108–122
Hasselrooth G (1944) Fan fare. Eugene Register-Guard. 27 April
Honey M (1980) The 'womanpower' campaign: advertising and recruitment propaganda during world war ii. Frontiers 6(1/2):50–56
Horrigan B (1938) Laugh with barney. Petrol Eng 9(6):8
Jack J (2009) Science on the home front: American women scientists in world war II. University of Illinois Press, Urbana-Champaign
Kennedy J (2017) Oil and gas industry. Mississippi Encyclopedia. http://mississippiencyclopedia.org/entries/oil-and-gas-industry/
Kentucky New Era (1949) Greer Garson weds Texan oil promoter. 16 July
Kitch C (2001) The girl on the magazine cover: the origins of visual stereotypes in American mass media. University of North Carolina Press, Chapel Hill

MacGregor MJ (1981) Integration of the armed forces 1940–1965. Center of Military History, Washington, DC
McEuen MA (2010) Making war, making women: femininity and duty on the American home front, 1941–1945. University of Georgia Press, Athens
Pidgeon ME (1946) Employment of women in the early postwar period: with background of pre-war and war data. United States Department of Labor, Women's Bureau, Washington, DC
Pitzulo C (2008) The battle for every man's bed: *playboy* and the fiery feminists. J Hist Sex 17(2):259–289
Pitzulo C (2011) Bachelors and bunnies: the sexual politics of playboy. University of Chicago Press, Chicago
Price GL (1997) Memorial to Dorothy Jung Echols. Geol Soc Am Memor 28:69–72
Rives WT (1941) Two girls achieve success as geologists in Texas. Saint Petersburg Times (29 April)
Santana MC (2016) From empowerment to domesticity: the case of Rosie the Riveter and the WWII campaign. Front Sociol 1:1–8
Steiner-Fleury B (2012) Disposable Heroes: the betrayal of African American veternas. Roman & Littlefield, New York
Temin P (2022) Never together: the economic history of segregated America. Cambridge University Press, New York
Troyan M (1999) A rose for Mrs. Miniver: the life of Greer Garson. University of Kentucky Press, Lexington
Turner TW (1946) Science teaching in Negro colleges. J Negro Educ 15(1):36–42
Warren W (2000) Black women scientists in the United States. University of Indiana Press, Bloomington
West M (2020) The birth of the pin-up girl: an American Social Phenomena, 1940–1946. Ph.D. Dissertation, University of Iowa

Chapter 5
Microscopes and the Post-War: "Women at Work"

5.1 Introduction: Women and Microscopes

After the end of World War II, women in geology faced another shift in identity, as they were increasingly becoming more active in the oil industry, but they were also more involved in the field, as well as feeling that they could advocate for their rights in their industry. This chapter seeks to examine the post-war experiences of Women within three different lenses: the transition in geology and oil exploration to the emphasis of micropaleontology, movements to fight for their rights in the oil industry, their transition back into peace-time society, and the industrial embrace of women serving the oil industry and geology in more technical and supplementary positions.

The combination of government promotion and newspaper amplification continued to make certain roles in petroleum geology advantageous to women, but women themselves probed their ability to work and succeed, despite previous assumption about women working. The image of the daring, petroleum geologist working in the field, as well as the technical expert working behind the microscope persisted and made space for working women, while also being influenced by ideas about gender and femininity (pictured in Fig. 5.1).

In the 1959 edition of *Careers for Women in the Physical Sciences*, in the section "*Women Geologists at Work*," there were estimates that 400 of the estimated 14,000 working geologists were women. And women since the war were likely excelling (Barber 1959, pg. 35–40). The publication proclaimed that "A number of these women have been in this occupation for some time and have achieved recognition for distinguished work." (Barber 1959, pg. 35). And though men could get jobs in the field, women often started their careers doing field work and had less access to different types of jobs. The example was given that men's opportunities, compared to women's opportunities that, "...men geologist can find entry jobs as field

E. A. Driggers, *Glamour and Geology*,
https://doi.org/10.1007/978-3-031-64525-9_5

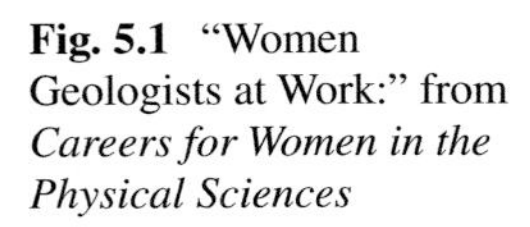
Fig. 5.1 "Women Geologists at Work:" from *Careers for Women in the Physical Sciences*

assistants, but few women geologists start their careers in this type of work." (Barber 1959, pg. 35).

By 1959, many of the assumptions about women's abilities to work in the field were dismissed, and no doubt likely came from the newspaper stories promoting the strength, but femininity, of women in the petroleum field. The sentence summed up a lot of the changes in attitudes toward women, "The women who have successful adjusted to the rigorous nature of the work, extensive travel, and irregular hours may serve to encourage employers to offer greater opportunities to women geologists in the future." (Barber 1959, pg. 35). And newspaper stories continued to emphasize a particular type of role for women both in geology and in the petroleum industry.

In a footnote, the publication even emphasized how much one woman enjoyed field work, especially to working in an office (Barber 1959, pg. 35). She went on walking around 10–15 miles daily on field work in Africa, which involved her traveling and strapping on fifty pounds of field equipment, and she also had to cook her food in the field. The publication promoted that, "A few women have established high reputations in field work for individual companies." (Barber 1959, pg. 35). And this unnamed Ph.D. female geologist had her work recognized in the publications of the American Association of Petroleum Geologists.

The unnamed female geologist also appeared in the *Leaflet* in 1959 that promoted women entering into the sciences (United States Women's Bureau 1959). In the section "Women Who Have Succeeded," there is a mention of an unnamed geologist in the collection of images, such as the one in Fig. 5.2 (United States Women's Bureau 1959, pg. 3).

A woman GEOLOGIST who is married and has a Ph. D. is a consultant and explores for oil for a large firm. Her most recent trip was in Africa.

Fig. 5.2 Cartoon of a female geologist from *The Leaflet*

The publication was put out exciting jobs and defined science with increasing excitement for women entering scientific, at both the expert and technical levels. *The Leaflet* promoted the future of science open for girls,

> When you take your first trip to the moon—thank the scientists and not your lucky stars!
>
> Scientists with their constant experimenting are responsible for much of the world's progress in many new fields.
>
> Today, there are not enough qualified scientists to meet the demand.
>
> Tomorrow's scientists will find rewarding opportunities in industry, in teaching, in government—
>
> [Sic]As researchers, technical writers, consultants, librarians [sic], supervisors, and field workers.
>
> *This is the time for more girls to consider careers in science.* (United States Women's Bureau 1959, pg. 2)

At the end of the publication, there was also an emphasis on nationalism. The publication ended with the heading "Invest in Your Future" (United States Women's Bureau 1959, pg. 9). Not only is job training, as well as education good for the individual, but it was good for the nation: "For the Nation, an adequate supply of trained workers is essential to economic development toward the goal of a fuller and richer life for all" (United States Women's Bureau 1959, pg. 9).

During the post-war era, there was a greater emphasis on science education because of the threat of Communism to the United States. The National Defense Education Act and other government bills wanted to mobilize students in colleges and university to pursue what we would now consider STEM education (Science, Technology, Engineering, and Mathematics) so that the United States could keep up with the USSR in the space race, especially after it had the first major success with the launch of the first man made artificial satellite, Sputnik (Leib 1999; Kallen 2019).[1] National interest promoted men, and later women, entering into the sciences, though not all races were advantaged by government programs (Reichstein 1999).

[1] See https://www.npr.org/2022/08/11/1116880026/stirred-a-debate-over-the-government-s-role-in-helping-pay-for-higher-education . Other scholarly studies of the origins of the United States Federal Student Loan system include some controversial studies like Shermer (2021).

In the post-war period, the United States Bureau of Labor Statistics studied the post-war labor landscape in *Manpower Resources in the Earth Sciences* published in 1954 (Shapiro 1954). The study found that women held an educational advantage over men, despite making up only three percent of the overall earth science workforce of meteorologists and geologists and with 1 percent of all geo-physicist being female. The study found that the average age for female geologists was 33 compared to 36 for men, and in geo-physics, it was 32 for women and 36 for men. Despite the small numbers, the difference in education was significant enough for the report to highlight, "The most marked difference in the educational background between men and women geologists and geophysicists was the higher proportion of women who held master's degrees." (Shapiro 1954). Only 1.1% of the women polled in the study did not have a degree in geology. However, the master's degrees for women in geology were few and far between at many institutions. So, there is much historical work to be done on where and what institutions were producing female geologists with master's degrees. For instance, at Texas Tech University, the first female geology master's degree was granted in 1939, and there was a gap in graduates until 1950.[2] Other institutions that historical had female geology students and faculty also produced a fair amount of female graduate students.

5.2 The Fossil Hunter with a Microscope

The 1959 *Careers for Women in the Physical Sciences* defined female geologists working in industry, in addition to clerical consulting, academic, research in the field, and writing, as working with the microscope. One type of job that is explicitly highlighted is the need for women trained in essentially micropaleontology and geologic work that utilized the microscope,

> One type of job open to women geologists in industry involves the microscopic examination of oil-well samples in the laboratories of petroleum companies. Women also perform a variety of chemical and physical tests to identify the different types of rocks and determine their age or composition. This identification helps to form the basis for decisions as to where the drill or tunnel for purposes of locating oil deposits and extending those already discovered or abandoning them when they appear to be unprofitable. Other duties may include evaluation of field observations, computation of probable locations of subsurface strata, and assistance in map preparation. (Barber 1959, pg. 36)

The success of women providing laboratory and microscope expertise continued the national and industrial interests of employing women and expanding the female workforce in oil discovery (O'Riley 1933).

In 1981, Jere H. Lipps attempted to define the field of micropaleontology in the entitled "What, if anything, is micropaleontology" (Lipps 1981). At the time of publication, Lipps pointed out the problems associated with defining

[2] On a research visit to Texas Tech I was given an internal list of all the graduates of the undergraduate and graduate degree program.

micropaleontology, it's hard to define and seemed to only involve geologists, as biologists did not associate themselves with the field. And the field was mostly at the service of geologists, and the term micropaleontology had been in use since the 1880s.

Micropaleontology is associated with the discovery of very small fossils that help locate patches or stores of oil. Lipps noted that fossils are markers that are important,

> …few micropaleontology papers in recent years are *really* concerned with the animals or plants as once living organisms. The fossils are merely convenient markers of ancient environments, currents, climates, or, most commonly, are tools for finding stratigraphic position. Very few studies actually deal with ecologic, biologic or evolutionary problems of the microfossils themselves. Micropaleontology has always been that way, and, as such, it has served geology exceedingly well! (Lipps 1981, pg. 169–170)

Technology improved the abilities of geologist looking for small signs and leads for oil in the field, and the need for experts with training in the use of microscopes opened up opportunities for women to work in geology (Gries 2018).

Earlier studies and examination of the history of the field, such as the 1959 study "Fifty Years of Micropaleontology," placed the origins of the field, especially its utility to oil discovery on the work of women (Howe 1959). Howe historicized that the "Credit for the revolution that took place in the use of foraminifera properly goes to the three ladies Esther Richards Applin, Alva C. Ellisor, and Hedwig T. Kniker. Their paper on the 'Subsurface stratigraphy of the coastal plain of Texas and Louisiana' (1925) had been presented before the Annual Meeting of the American Association of Petroleum Geologists in Shreveport, and the discussion which followed opened the eyes of oil company executives"[3] (Howe 1959, pg. 511–512). This drove the demand for women, or anyone, who could use micropaleontology to find oil. In examining the historical enrollment of college studies in 1941, there were estimated 500 persons practicing micropaleontology and working in oil. (Howe 1959, pg. 512)

In 1941, there was yet another retrospective historical study of micropaleontology by Carey Cronies entitled, "Micropaleontology—Past and Future" published in the *American Association of Petroleum Geologist Bulletin* (Cronies 1941; Sloss 1972). Cronies cites the origins of micropaleontology going all the way back to the eighteenth century and even older origins with the advent of the microscope with Anton Leeuwenhoek (Cronies 1941). By 1941, authors like Carey Cronies were accrediting women like Esther Applin, Hedwig Kniker, and Alva Ellisor as founding contributors to the application of micropaleontology to the discovery of oil (Cronies 1941, pg. 1221). The article cited the continuation of ideas and laboratory investigations that were going on in the early twentieth century. One of the geologists cited, Alva C. Ellisor, had her work cited at looking at cutting and her employment in Humble Oil company as important to finding wells. She was able to find micro-fossils at a low depth, as the Humble Oil Company had drilled about 3000 feet

[3] Esther Richards drops the Richards when she marries, and goes by Esther Applin.

and the appearance of micro-fossils indicated that there was oil. The fossils were then sent out to identification, and the results ultimately made it into a paper co-written with Dumble and Ester Richards entitled "Recent Geological Work in the Gul Coast Oil Fields"; that Richards read at the Geological Society Meeting of America in 1921.[4] Richards, Ellisor, and Kniker ultimately published a paper in 1925 entitled "Subsurface Stratigraphy of the Coastal Plain of Texas and Louisiana."[5] Cronies acknowledges that though he sketched out historical work by academics and comparing the prior work of others,

> Despite the development just described, the others to be mentioned later, and in spite of the of the fact that as early as 1919 C. R. Eckes used microscope examination of well cuttings to interpret logs of the Texas Duffer wells, *applied* micropaleontology really did not get well started in Texas Gulf Coast until 1920 when Alva C. Ellisor, Hedwig T. Kniker, and Miss Richards began their studies of Foraminifera. (Cronies 1941, pg. 1222)

However, the author pointed that the historical story has continued to focus on these three women starting off the oil revolution in micropaleontology. He also pointed out that though these women were significant in the development of micropaleontology, there might be other individuals that deserve some credit in its history, such as the interpretation of the fossils and contributions from companies such as Humble Oil, and Rio Bravo Company. He wrote that "It has become traditional that Mrs. Applin, Miss Ellisor and Miss Kniker are regarded as among the earliest pioneers in the micropaleontology of the Gulf Coast. But there will probably always be some questions as to whether or not the Humble laboratory in Houston was established as early as that of the Rio Bravo Company" (Cronies 1941, pg. 1223). But the author comes to the conclusion that these secondary claims might not be as important as the ongoing tradition as the three women contribution to the importance of the application of micropaleontology: "But Mr. Wallace Pratt might well have said, 'We were there, too'." (Cronies 1941, pg. 1223) Also, many of the most prominent women were trained by a geologist, Dr. Udden, who had an established lab in the Texas Coast, and trained women like Kniker, in micropaleontology, and she continued to contribute, with the other women, to its application. Women were, therefore, highly involved in micropaleontology and its usefulness to the oil businesses in Texas.

The Women's Bureau studying women involved in scientific work in the United States maintained many of their war jobs and found their utility in both geology and the oil industry through working in jobs that utilized micropaleontology, "…most of the women with graduate training in geology who were working in the oil industry in 1947 were engaged in the microscopic examination of rock cutting and cores from well borings, studying the lithology and faunal and floral contents of the rocks" (Zapoleon et al. 1948–1949, pg. 7–10). The publication went on to demark

[4]Also see Gries work on the subject, her write up is excellent: https://explorer.aapg.org/story/articleid/42216/three-women-one-breakthrough, and Gries (2018). See an excellent biography of Applin at: https://www.museumoftheearth.org/daring-to-dig/bio/applin.

[5]Please see Monica Kortsha's fantastic work that was published online in 2017: https://www.jsg.utexas.edu/news/2017/11/10881/. She also lists the title of the paper from 1925.

jobs in micropaleontology as appropriate for women, "Women employed in this work are often micropaleontologists. Many companies have found that women are well suited for this work not only in routine analysis but also in research" (Zapoleon et al. 1948–1949, pg. 7–10). Women were also working as "petrographers, stratigraphers, and geophysicists," and these jobs were only open to workers with advanced degrees (Zapoleon et al. 1948–1949, pg. 7–10).

5.3 The Application of Micropaleontology and the Microscope

The oilfields, and the offices and labs associated with oil exploration, became the perfect space of application of the new identification techniques of micropaleontology and opened up positions for women who were trained at using microscopes. Newspapers in the post-war era, or after the end of World War II, ran stories about women using scientific techniques to find or hunt for oil. Many of these articles romanticized the scientific nature of the women's work.

In 1947, a headline in the *Kentucky New Era* read "Lady Fossil Hunter is Aiding U. S. Oil Search." The article opened with the promises that micropaleontology held to the success of United States oil, "A little fossil 200 million years old many have led Americans to oil deposits which could help swing the scales to victory in case of war" (Kerr 1947). Though the article was published in 1947, looming and growing conflict with the Union of Soviet Socialist Republics (USSR and here on forward the Soviet Union) (Kerr 1947). Oil, and energy security, was seen as crucial to the battle of a capitalist nation like the United States, against a communist nation like the Soviet Union. The Cold War, an increasingly contentious relationship between the United States and the Soviet Union, would define conflict and diplomacy for much of the post-war era until the 1980s (Leib 1999; Kallen 2019; Leslie 1993; Reichstein 1999; Wolfe 2013; Oreskes and Krige 2014). The article framed the importance of oil in regards to scarcity and energy security. The article conveyed the imperative of finding oil, "The world is running short of cheap oil—vital to modern warfare and industry—and the great powers are on a constant hunt for it. Whether the little fossil found in the Peruvian Andes will lead to it, only time and drilling will tell" (Kerr 1947). The smallest information or fossil could lead to the biggest discovery of oil, especially in the scramble between potentially waring global powers.

The work of women would be of crucial importance to national defense and oil security. The article emphasizes the importance of women, "But whatever the results, a woman will have played a vital part in bringing them about" (Kerr 1947). The article then focuses on its subject, Valerie Newell, identifying her as the "wife of Dr. Norman Newell, curator at the American Museum of Natural History, New York, and professor of geology at Columbia University [,]" he was also teaching at Columbia University as well (Kerr 1947). Starting a biographical type article in this manner would not pass the Finkbeiner test that encourages female scientists to be defined independently of position their husband holds.[6] In this case, Valarie Newell was likely working in tangent with Norman Newell, as he was leading the expedition, and their working relationship went beyond simply their marriage.

The article goes on to praise Valarie Newell's strength in persisting through the elements to attend a strenuous expedition, "She climbed the chilly Peruvian Andes with the expedition he led in the search of fossils last summer, superintended supplies, boxed and shipped specimens and now, back at the museum, she is grinding the thin sections of fossils for microscope study" (Kerr 1947). Valarie Newell was a major player and contributor to the expedition.

Newell was an "invisible technician" and contributed a lot to Norman Newell's research.

The concept of invisible technicians is an important concept in the history of science. Historians of science like Steven Shapin have highlighted the importance of those working often in technical roles that contribute to major breakthroughs in science but do not get the credit that they deserve (Shapin 1989). Credit is denied because of their class or status in science community, as well as gender (Shapin 1989 and another example is Eads 1947). Other key examples, based on the gender of the knowledge creator, include Mileva Marić and Rosalind Franklin (Popović ed. 2003; Maddox 2013).

The American Museum of Natural History listed the expedition to the Andes of Peru.[7] The expeditions that the couple both traveled as members, yielded specimens. The Newells had been married for twenty years at the time of the writing of the article in 1947, and they met as undergraduates studying at the University of Kansas. The article expressed that Norman Newell was highly interested in paleontology and that interest fueled their academic trajectories, which were often together (Kerr 1947). However, further research by the author of this book could not find out whether Valerie Newell had any college level training in geology, and the author would argue that her high intelligence indicates her self-actualization and self-training in the subject.

Writing the backstory for Valerie Z. Newell, or Valerie Zickle prior to her marriage, revealed additional facets to an incredibly talented woman. Her training was not in geology, as she pursued journalism at the University of Kansas. She completed her Bachelor of Arts in 1922 and then her Master's of Arts Degree in Journalism in 1931 with a thesis entitled "Court Action as Influenced by Public Opinion Directed by the Newspaper" (Newell 1931).[8] The master's thesis examined how a prominent murder was portrayed in newspapers and the reporting's influence on the trial itself.

---

[6] For more information about writing about women in science see: https://www.nature.com/scitable/forums/women-in-science/writing-about-women-in-science-i-finkbeiner-102113387/

[7] There are archival records concerning the expedition, which are held at the American Museum of Natural History: https://data.library.amnh.org/archives-authorities/id/amnhc_2000449

[8] There was biographical information about Newell in her digitized thesis: https://kuscholarworks.ku.edu/handle/1808/23792

In 1927, the *Lawrence Daily Journal-World* published in their society section under personals that, "Miss Valerie Zirkle, who received her master's degree from the University this summer, will leave this afternoon for Oklahoma City. She will teach in the Oklahoma city high school this fall where she has accepted a position as head of the department journalism" (Lawrence Daily Journal-World 1927). According to Norman Newell's memorial, he and Valrier married in 1928, the next year after this newspaper reported on her plans. She likely traveled with Norman Nowell to the Yale University at some point and was hired by another faculty member at Yale, Charles Schuchert, "for temporary curatorial work," that Schuchert paid out of his own pocket during the Great Depression, and Norman completed his doctorate by 1933 (Kerr 1947; also see footnote about the biography of the Newells and their work at the American Museum of Natural History below).

The article goes on to imply that their interests overlapped, as both were interested in the Permian layer. Normal Newell studied the Permian layer as a professor and Valerie Newell studied the Permian layer by completing fossil discoveries and digs. The Permian layer is a part of the earth that is millions of years old, which spans from the Texas region to New Mexico. There is a lot of gas and oil stuck in this layer, even to this day, and oil exploration in this region made many oil and gas companies rich throughout the 1980s (Warner 2007).

Valerie Newell was an accomplished paleontologist and geologist, even though her academic status or publications did not indicate her diligent work. The article reported that, "While he was a university professor, he and his wife dug up Permian fossils in Oklahoma and Russia and, while he was oil consultant to the Peruvian government during World War II he saw signs which made him think he could find Permian fossils in the Peruvian Andres" (Kerr 1947). The two continued to work together on expedition that were funded by academic institutions, as well as American oil companies.

The article also made mention that the threat of oil insecurity from the United States was also at risk because of "Arabian oil" (Kerr 1947). Oil from the middle east was going to increase the supply of worldwide oil and open up new oil fields. The expertise of expert fossil explorers and scientists like Valerie Newell and Norman Newell was important to maintain the needed supply of oil, and it was expertise that all oil exploration hinged on, "Drilling oil wells, however, is extremely expensive and careful surveys of rocks must be made to determine favorable regions before a hole is sunk. Even then the financial risk is great. Nobody knows just what produces oil, but in the past it often has been found in rock strata containing Permian fossils. So When the Newell expedition discovered Permian fossils, it was good news to the oil backers" (Kerr 1947). The expertise of fossil seekers like the Newells paid off for American oil exploration in places like Peru.

The article ran in many newspapers in addition to the one mentioned in Kentucky. The article, in various forms, also ran in *The Deseret News* out of Utah, *The Edmonton Journal* out of Canada, and the *Prescot Evening Courier*, *The M'Alester News-Capital*, *The Marshall News Messenger*, and *The Enid Daily Eagle, John City Press, Long Beach Pres-Telegram*, and many other newspapers. However, *The Deseret News* stopped the article prior to the praise that Dr. Norman Newell gave to his wife and the article mentioned the serious contributions that Valerie Newell made to Norman Newell's publication (The Deseret News 1948).[9] While some of the publications emphasized the importance of Valerie Newell and her contributions to oil discovery, other the publication chose to emphasize the importance of her romantic relationship or omitted her contribution to science like *The Deseret News*. Other publications included her picture, like the one found in Fig. 5.3.

The article continued to discuss the work of Valerie Newell, as the *Kentucky New Era* did not end the article at the discussion of the oil company likely making money (Kerr 1947). The oil companies were most likely not going to look for oil in the mountains of Peru's Andes. The fossil find was an indication that there was likely an ancient sea filled with creatures which died and turned into oil after millions of years of being under pressure.

Fig. 5.3 Image of Valerie Newell (Edmonton Journal 1948)

[9] *The Deseret News* is owned by the Church of Jesus Christ of Latter-Day Saints; see its history in McLaws (1977).

> The significance of the Newell's Andean fossil finding to them is this: Millions of years ago a shallow temperate sea covered part of the western hemisphere from the spot where the Andres are now to the Amazon valley. Then, due to some unexplained disturbance of the earth, the western part of the sea bottom was lifted high in the air and formed the Andes mountains. In time the sea disappeared. Today the western part of the old sea bottom lies deeply buried under the Amazon jungle. (Kerr 1947)

The fossils that Valerie Newell found would help companies find where the Amazon jungle covers the ancient ocean, where there is likely oil in more accessible areas than the mountains.

The article then, presumably, reported Valerie Newell's explanation of the layers of the earth, as she analogized it to be a cake. The *Kentucky New Era* explained the old ocean's location through fossils and soil analysis,

> Thus the same Strata (the Permian) found in the Andes can be looked for under the Amazon jungle and there the oil companies persuamblye[sic] will search for oil. Moreover, due to the evolutionary process which staked the "vanilla, strawberry chocolate" layers in the same orderly sequences, they know when they have struck one layer whether they may expect to find the others looking above or below. In digging in the Amazon valley they will be working with at least a good gambler's chance of striking oil. (Kerr 1947)

The article then ended as if the reporter was continuing to talk to Valerie Newell, "As Mrs. Newell sorted the Peruvian fossils recently in preparation for her husband's research, Dr. Newell paused beside her and dropped one hand on her shoulder: 'She has been indispensable to us,' he said. 'We couldn't do without her'." (Kerr 1947)

Many of the other articles ran a similar ending. And though Norman Newell praised her contributions, she was not mentioned in the publications of the expedition, such as the 1949 *Geology of the Lake Titicaca Region, Peru and Boivia* or 1953 *Upper Paleozoic of Peru* (Newell 1949; Newell et al. 1953). Valerie Newell is mentioned only one two times in Norman D. Newell's memorial that was published by *The American Geologic Society* (Rigby 2006). The importance of her romantic relationship, as well as her novelty in a "Strange job," was highlighted in the April 20, 1948 article "These Gals Have World's Oddest Jobs" (Schurmacher 1948). Valerie Newell was featured among women working jobs that the paper would feel odd to the reader: "Worm Rancher, Human Cannon Ball Lead Amazing List" (Schurmacher 1948). Valerie Newell was discussed as "a fossil hunter" and brought to the attention of scientists and government workers studying employment in Washington, D.C., for the United States government.

The article, published in 1948, reported that there was twenty-nine percent of the workers in the United States were women. The article implied that many of the jobs occupied by women were very traditional such as "house worker," or "teacher," and it has made government workers apt to assume that women working fit in those traditional jobs, but the article makes the point that not all female workers fell into traditional jobs: "This has led to over-confidence on the part of the forms and paper work experts who assumed that all working gals could be labeled. They were wrong" (Schurmacher 1948). One of them profiled, along with an earthworm farmer, was Valerie Newell.

The article emphasized the romantic relationship as leading to Newell focusing on fossil collecting as a job. The sub-headline ran "LOVE MADE Mrs. Valerie Newell as fossil hunter. Just like that." The interview recounted Valerie Newell explaining how her future husband distracted her from her studies at Kansas, "'The night before mid-term exams he persuaded me to go to the movies,' she says. 'I told him I had to study, but he talked me into going. I got a C'" (Schurmacher 1948). It was unclear from the article if that was a good grade for Newell.

Studying about fossil became a type of hobby for Valerie Newell. She started to increase her knowledge after her marriage, "Nevertheless, despite the occasional temptation of the movie she boned up on fossils. Then she married the professor and they went fossil hunting together" (Schurmacher 1948). The article recounted the difficulty and adventure associated with Valerie's new job on collecting fossils, "Last summer they roamed the Andes by jeep and on mule back. Mrs. Newell was the only woman on this and many others of the professor's expedition" (Schurmacher 1948). In addition to gather fossils and performing scientific work, the article pointed out that she also completed crucial logistics work, "She superintends supplies and the packing and shipping of specimens."

In a call back to the World War II profiles of women geologists and pinup geologists of 1943, the publication asked Valerie Newell what she did after a long expedition. The article recorded the emphasis on feminine beauty and the duality of the technically informed and highly stylized woman of World War II, "WHAT DOES she do when she gets back from an expedition to some remote spot? 'I visit a beauty parlor, get our clothes washed and go to a picture show,' she says" (Schurmacher 1948). During and after the War, women of a variety of backgrounds were being drawn into petroleum geology, at first to serve the war effort, but often stayed in employment, and were profiled in trade presses and other media, showing how women, geology, and the petroleum industry had changed. Social ideas about beauty remained coupled with technical mastery, in profiles about the work of women (Dahl 2010; Carter 2021; McEuen 2010).

Other women in geology and petroleum were profiled in similar ways as Newell, like Doris Zeller (Florence Times 1955). Zeller was listed as a high-achieving female scientist in several United States newspaper. Zeller received her bachelor's and master's degrees at the University of Kansas and later earned a Ph.D. in geology. Zeller might have been the geologist represented in the cartoon in the *Leaflet*. She was on faculty of the geology department at the University of Kentucky from 1960 until 1987.[10]

The 1955 profile that ran in *The Florence Times*, published out of Florence, Alabama, listed Zeller as an award-winning young woman of the year, as pictured in Fig. 5.4 (Florence Times 1955). Her profile listed her as likely the only female consulting to a "major oil company." Clearly, this assertion is incorrect, as this book speaks the women who were working in the oil industry in various advising and

[10] Collection: Personal papers of Doris E. Zeller, Kenneth Spencer Research Library Archival Collections (ku.edu): https://archives.lib.ku.edu/repositories/3/resources/3281

Fig. 5.4 Doris Zeller (Florence Times 1955)

consulting capacities. The profile also lists her professional acknowledgments and awards as the American Association of Petroleum Geologists recognized her as having "'genuine contribution to the science of paleontology" (Florence Times 1955). The knowledge of small fossilized animals was listed as crucial to oil discovery: "Her special study is small animals that lived millions of years ago: they help determine where to drill for oil." She was also working with her husband, prospecting for oil in the Amazon and working for the Brazil prospecting for oil (Florence Times 1955). Her work in micropaleontology is continuing to be cited in the literature as recently as 2004 (Mamet and Belford 1968; Edgell 2004).

5.4 Heroic Women and the Search for Uranium

After the war, some women were put to work examining uranium. The headline ran in *The Evening Star*, published in Washington, D. C., that "Uranium, Not Gold, Is Today's Favorite Find: Modern Prospector Uses Tools of Atomic Age" (Miles 1950). The story, published in 1950, discussed the work of a Mrs. Helen L. Cannon who worked as a geologist, as well as "geo-chemical prospecting unit of the Geological Survey," and found that plants and other vegetation were good sources of revelation of uranium deposits. Cannon, "...has found although daises may not tell, many other plants will reveal the presences of minerals, including vital uranium in the soil." She thought that using plants to source out valuable metals, "Working on the relation of plants to mineral deposits was first begun through studies of the effect of accumulations of copper, lead and zinc on plants" (Miles 1950).

The article continued to promote Cannon's conception of the value of plants and geology to finding valuable metals on the earth, "'Work on the relation of plants to mineral deposits was first begun through studies of the effect of accumulation of copper, lead and zinc on plants,'...Eventually, she reported, all the major metals of economic importance will be studied in this way" (Miles 1950). She argued that plants were useful to those ends in three behaviors. The first is that if plants are present, there is a high concentration of minerals in the soil. She called these plans "indicator plants" and often do not exist in areas that have poor mineral concentrations. The second part of Cannon's plan was examining those plants that grow in areas of specific mineral concentrations. The last suggestion in Cannon's plan was that with the careful examination of plants in the region, scientist can identify physical changes that correlate to changes in mineral levels in the soil. An example of such changes Cannon offered was that surplus minerals like cooper, as well as zinc, cause trees in the areas to turn a yellowish color. (Miles 1950)

She was able to find high levels of cooper and zinc in New York. The discover occurred when the farmers reported that their plants would not grow in the soil after a peat bog was reclaimed and because of the high amount of zinc it immediate the growth of any plants. Later on, a fuller study found a high amount of lead and zinc present in the nearby soil. Studying plants in this way also helped prevent having edible food grown in potentially dangerous heavy metals. Of worry to the food supply were crops like carrots, with long roots, that would absorb the dangerous metals.

The article then shifted to Cannon's background. The article described her as "The daughter of a geologist," who then later dated and married another geologist. Many people in her family, including her brother in law, was also involved in geologist. However, she did not desire the career field of geology for her own children, as she commented that she hoped her daughters chose something else. Cannon stated that, "Fascinating as she finds her field, she believes that two braces of geologists are quite enough for one family, and hopes her daughters, now 10 and 13 years old, will choose different careers" (Miles 1950).

Cannon, however, was a first in her field, as she was "….one of the first woman geologists to be hired by the survey during the war" (Miles 1950). Before she undertook wartime work with geology and plants, she had a career in petroleum geology. Cannon also remarked that she found petroleum geology a good place for women to work, especially in the field of geology, "Prior to her marriage, she worked for a large oil refining company, the area of her profession which she feels is probably the most satisfactory for women" (Miles 1950). Cannon was involved with micropaleontology in her work prior to World War II. The article summed up her contributions in microscopic paleontology but also related it to her experience as a woman, "In this work, a microscopic study of rock cuttings is used to determine the age of the rock being drilled, which in turn is related to the likelihood of oil deposits there" (Miles 1950).

She then commented on the field work component of petroleum geology, "Field work is a handicap for women in geology, she believes" (Miles 1950). She went on to continue to discuss the challenges of women in the field: "Among its difficulties is the fact that miners are superstitious about women in the mines and sometimes refuse to admit them, feeling that their presence there spells disaster." Interestingly, the article makes the analogy to the magical by the inclusion of spell, a pun.

Cannon then lamented how the inability to access mines and field sites made her job incredibly difficult, but yet she completed the work, "Mrs. Cannon has surmounted the disadvantages of field work herself, however, and has pursued her profession in Utah, Arizona, Colorado, Wisconsin and Wyoming, as well as New York" (Miles 1950). Many of those states likely were sites of petroleum geology field work. The article then went on to briefly explain her biography, discussing her bachelor's studies in geology at Cornell and her continuation of studies at Northwestern and Pittsburg, noting that her father was a geology professor at Pittsburg. She also received further training at the University of Oklahoma (Miles 1950).

According to her obituary published by the geological society of America in 1996, she earned a masters at the University of Pittsburg in the 1934 and then she worked for the Gulf Oil Company around 1935. Around 1938 she was employed by the US Geological Survey. The obituary does not make any direct mention of her work, but it emphasizes a paper published in 1952 about plants being useful in detecting heavy metals, as described in the newspaper article (McCarthy 1997). Some newspapers ran articles where she was stating that uranium deposits could be causing cancer cases to rise.

The article "Chances of Cancer, Other Diseases Depend in Part on Nature of Soil" featured Cannon who was working in 1964 at the United States Geological Survey, based in Denver, Colorado. She articulated the danger of soils to disease, "…[S]howing that plants and soils in areas with a high incidence of cancer and heart disease contain more arsenic, nitrate, boron, manganese, chromium, titanium and lead than in areas with a low incidence of the diseases. Tuberculosis was more prevent in the low-metal areas than in the high metal localities" (Saskatoon Star-Phoenix 1964). Her work, according to the obituary writer Howard McCarthy, was useful in finding large amounts of uranium in the states of the Western United States, such as Colorado, New Mexico, and Utah. McCarthy referred to the period of the 1950–1960s as the "uranium boom," because of Cannon's research and work (McCarthy 1997; Saskatoon Star-Phoenix 1964).

In the 1955, *The Times-News*, published out of North Carolina, ran a story that asked "Geiger Counter Obsolete? Weeds Can Count, Too" (Smith 1955). The story ran on the "Young Folks" section and was designed for children and young adult readers. Helen Cannon was credited with the idea, while working for the United States Geological Survey. She ran experiments with plants to see if they indicated Uranium levels. These experiments occurred in Colorado and New Mexica and were done for the Atomic Energy Commission. The plants exposed to uranium would show it physically, either dying or wilting. Cannon found that plants exposed to uranium grow "the prince's plum" and can be tested by being burned and their remains were analyzed in the laboratory.

The article went to explain that what Cannon was doing was a new way to discover energy and minerals: "…geobotany or geobotanical prospecting" (Smith 1955). The image in Fig. 5.5 was likely meant to get the readers excited and advertised the questions of a new science, perhaps to inspire young readers to study science: "It is a new science and much needs to be learned about it" (Smith 1955). The article also discussed how the Geiger counter worked to reveal the levels of uranium but was limited in comparison to plants. Though Geiger counters can only peer into

WHAT DO THESE TWO pictures have in common? Nothing? Guess again. They're both prospectors—old style and new. The Geiger counter, on the left, is thought by most people to be the last word in prospecting apparatus. But that little old nicotiana plant (right) has been finding—and mining—copper for years.

Fig. 5.5 "Geiger Counter Obsolete? Weeds Can Count, Too" (Smith 1955)

the soil to a degree of feet, plants have roots that tell about the soil contents a hundred feet below the grown and are far less expensive than purchasing a Geiger counter. And in fascinating factual information for young readers, the article continued to promote the virtues of Cannon's ideas: "Meanwhile plants continue to 'mine' many important elements. One of them is the selenium which the vetch can extract from soil c[ontaining] such small amounts that man has not yet devised a means of recovering them from the earth" (Smith 1955).

5.5 Women, Narratives, and the Microscope: Jane Gray (1929–2000)

Women's scientific work in the 1960 included descriptions of the physical appearance of the female researcher, such as the story of the paleobotanist Jane Gray, pictured in Fig. 5.6. The article featured her work with micro-fossils but described her physicality in working in her lab. The article appearing in the *Eugene-Register-Guard* in 1965 ran with the headline "Research Analyst's Job Nothing to Sneeze About." The description in the article described Gray's research, "Most people recognize pollen to be that stuff hay fever sufferers complain about, or move away from, or sneeze long with." Then Gray's physical work was then described, "But to the dark-haired women bent over her high-powered microscope, the minute pollen grain holds the key to plaint life that existed in the Pacific Northwest more than one million years ago" (Rakish 1965).

The article went on to discuss the nature of women in micro-botany. Microbotany uses very small pollen samples and allows the scientists to understand plants and the environment life that existed millions of years ago. The diligent scientists written about the article worked 10 hours a day and have recorded microscopic pictures of around five thousand samples. And while the article praises her technical expertise and hard work ethic, it then discusses her perseverance and physicality. The article,

Fig. 5.6 Dr. Jane Gray (Rakish 1965)

appearing in 1965, has the same style of those written in 1943. Gray was described in the article as,

> This tall, little woman hardly seems the type to go "out-in-the-field" equipped with hammer and shovel for obtaining samples from sedimentary deposits. But that's another demanding aspect of her job. These samples must be carefully cleaned and placed in plastic bags to avoid contamination from present day, free-floating pollen." (Rakish 1965)

Micropollen analysis is a type of micropaleontology practice (Moore Jr. and Echols 1979). Gray makes many transitions from the field to the laboratory, much like the other geologists discussed in this chapter. Gray received her Ph.D. in geology at the University of California and went through additional training and employment at the University of Arizona. She took her current job in 1961 where she works in the museum and research at the Museum of Natural History on the campus of the University of Oregon.

Then, like the profiles in oil trade journals, the article focused on her demurring her physical beauty and her experiences being a female scientist (Rakish 1965). She comes from a military service background ("'Army-brat'"), loves her cat, and also reads and listens to music. Gray also rails against stereotypes of women working in science, "She doesn't think that 'being a brain' is detrimental to a young girl of a woman. She has definite ideas about women who use feminine wiles to get ahead in the business world" (Rakish 1965). She dismissed beauty and seduction to rise in the professions: "'If a woman is good in her profession, feminine wiles are always a welcomed asset,' Miss Gray said. 'But all the feminine tricks she can think of won't help her if she can't carry out her daily responsibilities'" (Rakish 1965).

Like many of the other geologists that appeared in newspapers in the 1940s, she stressed her ability to do her job, despite prejudice, and not having to rely on beauty, and like the women during the war, continually dismissed it. The reporter asked Gray if she experienced any professional struggles due to prejudice. She responded that though there was prejudice, a woman could overcome it with competence and technical mastery: "'In any field in which men have been the prime motivating force, there's a bound to be some prejudices. It's hard to generalize, but I believe it depends in part upon the woman. The more professionally competent she is the more sentiment she's bound to encounter" (Rakish 1965). The best women, she argues, simply overcome prejudice: "On the other hand, competence is always respected in any field and she'll over come resentment. After all, a really good woman won't be held back just because she happens to be a woman" (Rakish 1965).

In her memorial published in *Nature*, her colleagues remembered her as "...intolerant of misbehavior professional or otherwise..." (Shear 2000). Looking retrospectively in 2000, William A. Shear wrote that she opened doors for women in science but also hindered her personal relationships,

> The 1950s and 1960s were not particularly female-friendly times in science, and Gray's uncompromising personality and readiness to defend her work made her rise through the academic ranks difficult. 'Anti-nepotism' regulations made it impossible for Jane and her husband, Antone Jacobson, to work at the same institution. Long periods of separation eventually led to a divorce. (Shear 2000)

The memorial also discussed Gray as a controversial figure, "Jane Gray, a controversial figure in paleoecology, died of acute liver failure caused by cancer on 9 January 2000" (Shear 2000). But the memorial did not explain. Upon further research, Gray's work in the women's rights movement appeared in newspapers.

Gray became more active in advocating for women working in science the next decade after the article discussing her paleobotany work. In 1975 an article featuring comments from Gray carried the headline, "Court Rules State Unfair to Women" (Eugene Register-Guard 1975). This article features a court case where women like Gray started to transition from the narrative of the 1940s–1960s that women could simply transverse or overcome oppression and prejudice by being superior scientists. She, and co-worker Carmella Hartin, sued their employer the State of Oregon, and the state court system found that Oregon had "...discriminated against women by using different life expectancy tables for men and women in calculating benefits for retired employees under the state retirement system" (Eugene Register-Guard 1975). Gray heighted the importance and stakes of the ruling, "'It's been a long fight and I see the ruling as one, a case of forcing social change, and two, a case for women everywhere,' she said. 'There are other actions like this in other states, and I think the ruling will have some effect there'" (Eugene Register-Guard 1975).

In 1970, Gray was involved in a study of female faculty at the University of Oregon that was reported in the article's title as "OU women faculty separate but unequal" (Anderson 1970). The report, authored by Joan Acker of sociology, Jane Gray "curator of paleobotany," and Joyce Mitchell of political science, pointed that a major problem, beyond lack of discrimination and social exclusion, was the lack of female mentorship: "'With so few available role models of working women professional, potential women scholars may see no realistic future in an academic career.["] (Anderson 1970).

In 1982, another, similar story appeared in the *Eugene Register-Guard*, which profiled Gray and her work with the microscope. The later article omitted much of the description of Gray's physical features and there was no discussion of her experiences as being a woman in science (Stahlberg 1982). The article only focuses on Gray's scientific work. But in 1982 Gray had transitioned to working with oil drilling. In order to get deeper samples of pollen, she was working with core samples that came from oil wells in Libya (Stahlberg 1982). The samples from Libya produced very old fossils of plants, showing that ancient oceans of 600–800 million years ago flourished in the Cambrian period (Stahlberg 1982). The article implied that Gray was "very controversial" because of her theories about the age of plant life and was preparing to defend her ideas: "'It is going to be quite a battle' among paleontologists, she says" (Stahlberg 1982). Other articles discussing her work and high accomplishments in the study of ancient plant life omit any discussion of her experiences in science as being a woman, such as a report of her election in American Association for the Advancement of Science fellow in 1987 (The Bulletin 1987).

The portrayal of women in petroleum geology and geology often highlighted their work with the microscope. However, the narrative of the beautiful, yet technically knowledgeable woman doing detailed scientific work, despite her physical appearance, persisted from the 1940s into the 1960s. However, the narrative departures of the 1970s and 1980s will be discussed in Chaps. 6 and 7.

References

(1927) Personals. Lawrence Daily Journal-World. 31 August
(1948) Lady fossil hunters seeks oil source for United States. The Deseret News. 1 January
(1955) Twelve-year-old named one of "Ten young women of the year." Florence Times. 28 December
(1964) Chances of cancer, other diseases depend in part on nature of soil. Saskatoon Star-Phoenix. 28 December
(1975) Court rules state unfair to women. Eugene Register-Guard. 11 December
(1987) Pair of U. O. biologists honored for research. The Bulletin. 24 May
Anderson C (1970) UO women faculty separate but unequal. Eugene-Register-Guard. 22 October
Barber MS (1959) Careers for women in the physical sciences. Women's Bureau, Washington, DC
Carter IS (2021) The red menace: how lipstick changed the face of American history. Prometheus Books, Guilford
Cronies C (1941) Micropaleontology—past and future. Am Assoc Petrol Geol Bull 25(7):1208–1255
Dahl HJ (2010) Fearless and fit: American women of the cold war. Master of Arts Thesis. University of New Mexico
Eads J (1947) Washington social whirl described. Prescott Evening Courier. 29 December
Edgell SH (2004) Upper Devonian and lower carboniferous foraminifera from the Canning Basin, western Australia. Micropaleontology 50(1):1–26
Gries RR (2018) How female geologists were written out of history: the miscropelonotlogy breathrough. In: Johnson BA (ed) Women and geology: who are we, where have we come from, and where are we going. Geological Society of America, Boulder
Howe HV (1959) Fifty years of micropaleontology. J Paleontol 33(3):511–517
Kerr A (1947) Lady fossil hunters is aiding U.S. oil search. Kentucky New Era. 31 December
Leib K (1999) International competition and ideology in U.S. space policy. Int Stud Notes 24(3):30–45
Lipps JH (1981) What, if anything, is micropaleontology. Paleobiology 7(2):167–199
Kallen S (2019) Nationalism, ideology, and the cold war space race. Constellations 10(2). https://doi.org/10.29173/cons29377
Leslie SW (1993) The cold war and American science: the military-industrial-academic complex at MIT and Stanford. Columbia University Press, New York
Maddox B (2013) Rosalind Franklin: the dark lady of DNA. Perennial, New York
McCarthy H (1997) Memorial to Helen l. Cannon 1911-1996. Geol Soc Am Memor 28:53–55
Mamet BL, Belford DJ (1968) Carboniferous foraminifera, Bonaparte gulf basin, northwestern Australia. Micropaleontology 14(3):339–347
Miles B (1950) Uranium, not gold, is today's favorite find: modern prospector uses tools of atomic age. The Evening Star. 24 April.
Newell VZ (1931) Court action as influenced by public opinion directed by the newspaper. M. A. Thesis, University of Kansas
Newell ND (1949) Geology of the Lake Titicaca region, Peru and Bolivia. Geological Society of America, New York
Newell ND, Chronic J, Roberts TG (1953) Upper Paleozoic of Peru. Geological Society of America, New York
O'Reilly P (1933) Buy microscope instead of auto. Spokane Daily Chronicle. 19 January
Oreskes N, Krige J (eds) (2014) Science and technology in the global cold war. MIT Press, Cambridge

McLaws MB (1977) Spokesman for the Kingdom: early Mormon Journalism and the Desert News, 1890–1989
McEuen, MA (2010). Making war, making women: femininity and duty on the american home front, 1941–1945. Athens: University of Georgia Press.
Moore TC Jr, Echols JR (1979) Micropaleontology. In: Paleontology: encyclopedia of earth science. Springer, New York, pp 470–475
Popović M (ed) (2003) In Albert's shadow: the life and letters of Mileva Marić, Estein's first wife. The Johns Hopkins University Press, Baltimore
Rakish J (1965) Research analyst's job nothing to sneeze about: paleobotanist, Jane Gray, notes latest fossil tabulations. Eugene Register-Guard. 3 March
Reichstein A (1999) Space—the last cold war frontier? Amerikastudien 44(1):113–136
Rigby JK (2006) Memorial to Norman Dennis Newell (1909-2005). Geol Soc Am Memor 35. https://rock.geosociety.org/net/documents/gsa/memorials/v35/Newell.pdf
Shapin S (1989) Invisible technician. Am Sci 77(6):554–563
Shermer ET (2021) Indentured students: how government-guaranteed loans left generations drowning in college debt. Harvard University Press, Cambridge
Schurmacher EC (1948) These gals have world's oddest jobs. St. Louis Globe-Democrat. 20 April
Shapiro S (1954) Manpower resources in the earth sciences. National Science Foundation, Washington, DC
Shear WA (2000) Obituary: Jane Gray (1929-2000). Nature 405(34). https://doi.org/10.1038/35011189
Sloss LL (1972) Memorial to carey cronies. Geological Society of America Memorials. https://rock.geosociety.org/net/documents/gsa/memorials/v04/Croneis-C.pdf
Smith I (1955) Geiger counter obsolete? Weeds can count, too. The Times-News. 12 May
Stahlberg M (1982) U of o scientist turns clocks back: plants lived on land earlier than we think. Eugene-Register-Guard. 22 April
United States Women's Bureau (1959) Leaflet 32: science futures for girls. Women's Bureau, Washington, DC
Warner CA (2007) Texas oil & gas since 1543. Copano Bay Press, Corpus Christi
Wolfe AJ (2013) Competing with the soviets: science, technology, and the state in cold war America. Johns Hopkins University Press, Baltimore
Zapoleon M, Goodman EK, Brilla MH (1948–1949) The outlook for women in science. Government Printing Office, Washington, DC

Chapter 6
The Pressure of Doing It All: "Glamour Girls," Imagination, and Identity in Post-War Petroleum Industries

6.1 Oil Companies, Female Workers, and Magazines

Oil companies were active in recruiting women into all aspects of the oil business. A targeted ad often appeared in *Time* magazine in the 1950s. In the May 9, 1953, issue of *Life*, the ad "Man Hunt: Oil Companies Complete for the Class of '53," one of the examples rising college graduates was Commerce major Mary Uehling. The advertisement promoted that Uhleing "…will join thousands of young women in the oil industry" (Life 1953, pg. 150). The ad promoted the idea that each oil company was under pressure to compete for graduates to fill its employment needs. In addition to technical science jobs, companies wanted women working in geology in the field, as well as in non-field based jobs like: administrators, office employees, technicians, and civic speakers. Oil companies recruiting women created a world that promoted petroleum, but also a world with new opportunities for women.

Ads in *Life* magazine had a lot weight, as *Life* had a major circulation and was read by around ten percent of the US population during the post-war era. Many of the ads, produced by the American Petroleum Institute, featured women doing highly technical jobs and emphasized how many female employees made up the American Petroleum industry. In 1951, the ad in *Life* mentioned that there were around 15,000 men and women employed in oil industry constantly improving the production and efficiency of oil (Life 1951). Ads promoted the work of women in petroleum industry, as pictured in Fig. 6.1.

Oil companies were also interested in women, regardless of their age. The pamphlet put out by the Women's Bureau entitled "'Older' Women as Office Workers" encouraged women to seek out pilot programs that got women into the industry as "older" women. The organization featured was the Desk and Derrick Club (Ravner 1953). Schools and proprietary schools took advantage of the demand for women workers in the petroleum industry. Women, as well as men, are encouraged to enter the oil business.

E. A. Driggers, *Glamour and Geology*,
https://doi.org/10.1007/978-3-031-64525-9_6

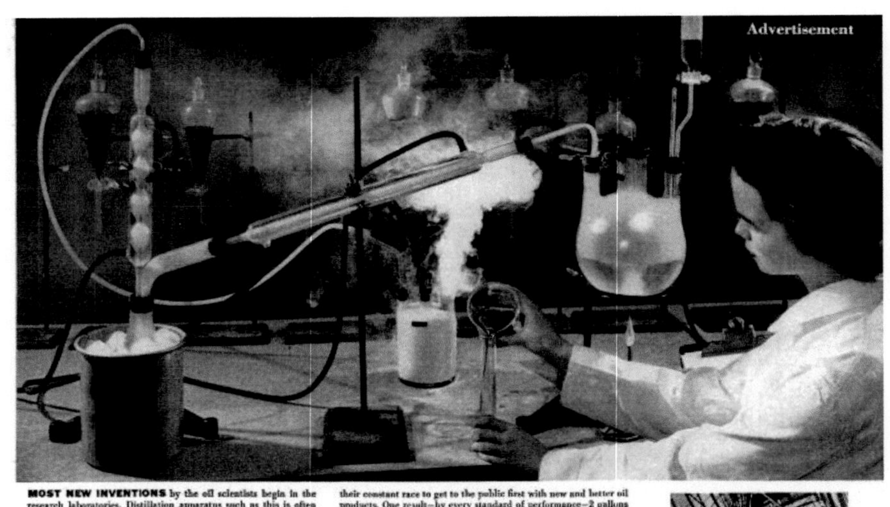

MOST NEW INVENTIONS by the oil scientists begin in the research laboratories. Distillation apparatus such as this is often used. U. S. oil companies employ more than 15,000 men and women research workers, spend more than 100 million dollars a year in their constant race to get to the public first with new and better oil products. One result—by every standard of performance—2 gallons of gasoline now do the work 3 gallons did in 1925, though today's gasoline is priced about the same—only taxes are higher.

8,179 New Oil Inventions Patented By Scientists In Five Years

Fig. 6.1 Women's work in oil was actively promoted in magazines like *Life* (Life 1951)

In 1955, *The Poca City News* ran a story explaining educational opportunities for students at Northern Oklahoma Junior College. The "Oil Educational Institute" was run for two days, which explained to both students and the general public, the process of extracting petroleum and manufacturing its byproducts. The article went on to record the director of a locally based oil company, David Tver, who promoted the opportunities for men and women in oil, "There is a great opportunity for both men and women to work in the oil industry in the U. S. and abroad, he said." (The Ponca City News 1955).

Other oil companies like Standard Oil Company of Indiana touted how many women not only worked at the company but owned stock and profited off the company. The eye-catching headline "WE SEND MONEY TO 40,000 WOMEN" ran in several different publications from 1949 to 1950 (Standard Oil Company 1950). The article promoted that exactly 41, 458 women owned some degree of shares in Standard Oil and the company was more than happy to send its investors profits on their shares. The ad went on to feature the diverse group of women who had received checks from the company, including Esther Anderson. Anderson was one of the 96,800 employees who owned stock in the company. The advertisement went on to explain that employees bought stock "voluntarily." The ad also tried to cater to married and disowned women as it would be a secure investment and allows women, like Mrs. V. E. Webb "…to maintain her home and to follow her hobby of fancy needlework." (Standard Oil Company 1950). Other oil executives in the 1960s described having "'oil running out of its ears," but they needed highly educated workers (Dawson 1961). Articles like "Industry Woods Students With Science Skills: Big Firms Scan Education Facilities in Locating Plants" advertising for scientific specialists promoted opportunities for women as well (Toledo Blade 1961). Companies had to offer high starting salaries to both men and women in highly specialized and technical skills.

In 1960, optimism about women entering the field of geology appeared in the young women's magazine *Seventeen*. In the sixth issue of 1960, the magazine ran a story entitled "Looking Ahead to College," where a professor of education, Dr. Charles A. Bucher, wrote promoting junior colleges, while also including a discussion of careers open to junior college graduates (Bucher 1960, pg. 134–135). He discussed opportunities in the clergy, economics, and architecture, the field of geology and its career outlook: "Stiff competition for those with only bachelor degrees. A bright future for those with advanced work, due to the expansion of the overseas petroleum industry" (Bucher 1960, pg. 134–135). Petroleum seemed to be a field gradually opening to women in the field of geology.

Newspaper ads featured women returning to the workforce in oil in 1962, such as "A Feminine Wheel in Oil" (Toledo Blade 1962). "Mrs. Charles R. Foust" who was both a wife and mother was a high-ranking administrator at Ohio Oil Company. The article noted that there were not many female executives, managers, and administrators in the oil industry. Mrs. Foust, as she is only identified in the article, was in charge of outreach and public relations with women, as she served as the director. The program, according to the article, was exceptional, "Mrs. Foust directs the Ohio Oil Co.'s program designed to interest women and at the same time inform them about the oil industry. In directing such as a program, itself rare in the industry, she has become one of the very few women in the field with such responsibility" (Toledo Blade 1962). Other colleagues mentioned how rare such an outreach program for women was among oil companies.

Foust's role was to facilitate educational speakers and program for women's social groups and community organizations. She had a public relation's background and trained featured speakers that would then go to groups and make presentations about the petroleum industry. In terms of demographics, there were 17 speakers out of the total 34 speakers that were women. Foust gave weekly presentations on "what industry expects of young employees" but also the advantages of taking children on trips. She noted that the largest problem of her job was "…rich disserts" (Toledo Blade 1962).

Foust's joke was likely an attempt to make her more approachable to social groups and make her an identifiable figure in the culture of the women she served. Diet culture in the 1960s, also called "reducing," emphasized restriction in eating, especially regarding sweets. Some diets of the time often only allowed women to drink a whole bottle of wine during the day, or just coffee, eggs, and steak. Many women developed lifelong disordered eating from such diets or cultural emphasis on women "reducing" (Wdowik 2017). The unrealistic body standards, physical illness that resulted from "reducing," and fat-phobia can be found in the changing body image of women, historicized in the book *Reducing Bodies: Mass Culture and the Female Figure in Postwar America*.

Beauty consultants, who presented at groups for women affiliated with the oil industry, promoted aspects of diet culture (Wdowik 2017; Matelski 2017; Doan 1961). An example can be found in the article "Beauty consultant gives sage advice." Miss Ruth Langford, a Canadian beauty consultant and "cosmetician," promoted elimination diets. She spoke at a meeting of the Desk and Derrick club, a group that promoted the oil industry to women, which will be discussed later in this chapter. She reminded readers in the article against the dangers of crash diets but promoted elimination diets: "Miss Langford emphasized that crash diets are bad for the body. All that is needed to be done is to cut out sugar, breads, salt and butter, and to exercise" (Doan 1961). She also promoted makeup and beauty culture: "'It is not every woman's privilege to be born beautiful but there is not a woman alive that cannot be made pleasant to look at' is a well known beauty truth." (Doan 1961) Makeup, a product produced by the oil industry, was integrated into the promotion of the oil industry in general to woman who attend Desk and Derrick. The women at the meeting received makeup and "skin" treatments under the advice of Ruth Langford. Women received the message that beauty culture improved their physical appearance and were pressured to pursue products and diets to become "more beautiful" in the 1960s and oil companies promoted those ideas.

Foust shared her own employment history that would be likely very similar to her audiences. Foust had children and left her job and then returned to work in the late 1950s (Toledo Blade 1962). After returning to work and taking a job at Ohio Oil, she started as a secretary, worked as a librarian, and continued to work up the Ohio Oil corporate ladder. Civic outreach from oil companies employing women allowed them access to educated and specialized workers, and these groups also spurred educational development and opportunities for women. The oil company portrayed itself as a place that was open to women and continued to voice the narrative of the exceptional women. The duality of the few and proud female employees were continued, while the oil companies advocated open spaces for women. The importance of women working in civic organizations and advocating for the importance and opportunities the petroleum industry could provide was also useful for women seeking employment and opportunities themselves. This was best represented in the organization Desk and Derrick.

6.2 Women Within and Connected to the Industry: Desk and Derrick

Sarah Stanford-McIntyre has published on the Desk and Derrick Clubs that were associated with the oil industry in the post-war era (Stanford-McIntyre 2022). Stanford-McIntyre's article posit that the work of women in both clerical and support helped the petroleum industry as a whole more fully develop

(Stanford-McIntyre 2022, pg. 6). The article also challenges assumptions about strong men in the oil industry that simply worked independently and generated their own wealth. Stanford-McIntyre challenges that in writing about the women working in the offices, "…it demonstrates that despite the industry's mythology of individual inventors and lucky wildcatters, oil was remarkably similar to other large-scale scientific and engineering enterprises during the middle decades of the twentieth century" (Stanford-McIntyre 2022, pg. 6). This story continues to be popular, as recent biographies of George Bush Sr. emphasize his fierce independence

The story of Desk and Derrick also shows the role of women in labor relationships and strikes: "As in the aerospace and nuclear industries, women were fundamental to oil industry technological developments and labor conflicts[5]" (Stanford-McIntyre 2022, pg. 6). The work of the women in the club helped with the industry oil in terms of re-branding as the article discussed labor conflict, and the culture of sexual harassment is also examined in the article,

> The club's vocal emphasis on scientific education and credentialization represented a bid for female inclusion within an increasingly technically complex professional world. However, entrenched workplace sexism and union hostility to changing labor structures limited member opportunities. As a result, Desk and Derrick's middle-class aspirations ultimately allied the club with industry rebranding efforts and helped support industry automation and union-busting. (Stanford-McIntyre 2022, pg. 7)

In this chapter, I want to focus on how women on the clerical oil industry found themselves in a high-pressured duality, of maintaining industry branding, while having to conform to standards of femininity.

Even working professional women, such as petroleum geology, found, however, an opportunity to make new rights for themselves, as the oil industry started to more focus on their role within the industry beyond propaganda and catchy newspaper images. This is especially true in the "Women's Focus Group" that was formed by female employees at Mobile Oil in the 1960s. There were technical speakers that came in to continue the professional development of the female workers, but contained within the files were also the protests to wear what they pleased in the office, which included mini-skirts and other new fashions of the 1950s and 1960s. Through collective action, these women re-defined the office space.[1]

In many newspaper entries, women leaders of Desk and Derrick often highlighted its mission and purpose. That purpose can be summed up in Helen Wilkie when she addressed local Canadian Desk and Derrick Club, claiming that the oil industries, like the two countries, shared a lot in common (Leader-Post 1959). And Wilkie, in giving the address of the club, framed the purpose of Desk and Derrick to empower women, "To promote among the women and employed in the petroleum and allied industry, through informative and instructional programs, a clearer understanding of the industry which they serve, to the end that the enlightenment gained thereby may increase their interest and enlarge their scope of service" (Leader-Post 1959). Desk and Derrick sought to engage women and women associated with the

[1] For an example of the women's focus groups see the entry in *Mobile World*, September of 1962.

oil industry. And she saw one of its point of engagement as public relations and described the club's activities, "The members had taken trips to refineries, oil fields and had witnessed seismic crews as work. 'I for one, now know that a "pig" doesn't have to have four legs, a "Christmas tree" can be put out in the field in the middle of summer and a "coke drum" at a refinery isn't a refreshment stand,' she said humorously" (Leader-Post 1959).[2]

She continued to describe the values and mission of Desk and Derrick, "Our motto is 'greater knowing, greater service'," Helen stated. "We automatically contribute as individual public relations representative for our companies, and in a larger for our industry" (Leader-Post 1959). Desk and Derrick attempted to add charm with an aggressive public relations campaign to get more access for women into the oil industry and open doors for career opportunities for them as well. Many women made a name for themselves in working in petroleum geology behind the scenes in public relations and administration.

In the *Kappa Alpha Theta Journal* from 1954, the stories of Virginia Dupies appeared with the headline: "Virginia Dupies Carves Career in Oil Industry" (Kappa Alpha Theta Journal 1954, pg. 31–32). Dupies lived from 1912 to 2007.[3] In the biographical write up of the Beta member from Southern Methodist University, the article reported that Dupies career work had been profiled in an earlier 1953 edition of *Dallas* magazine (Thompson 1953, pg. 63). She was also the president of the alumni chapter based in Dallas, as well as in other service capacities. Currently she was servicing as an executive secretary to one of the vice presidents in the Continental Supply Company. She was known for her creativeness such as fashioning a Christmas tree out of oil well parts. However, the article went on to describe her extensive knowledge of oil drilling and related equipment. Dupies received her degree in journalism as well as taught journalism at Southern Methodist. She had a reputation as being a solid communicator regarding oil drilling technology. In the original *Dallas* article, a picture of Dupies was included (Thompson 1953, pg. 63). Her picture, included as Fig. 6.2, was included in the profile with an oil derrick.

Dupies had many jobs at the Continental Supply Company, as the article described her as multi-talented and well versed in outreach organization like the Dallas Petroleum Club, even serving as its secretary. Dupies was highly engaged in outreach to the larger petroleum community:

> Although she denies emphatically that she is the "club woman" type, Miss Dupies early realized the importance to the oil industry of the new national association of Desk and Derick Clubs, composed of women in the industry. So she became a charter member when the Dallas club was organized two years ago, and is now completing her term as second president of the group. (Kappa Alpha Theta Journal 1954, pg. 31; Thompson 1953, pg. 63)

[2] I quote the newspaper article and the article is also directly quoting members. I used quotation marks to the best of my ability.

[3] Her obituary is still available to read: https://obits.dallasnews.com/us/obituaries/dallasmorningnews/name/mary-dupies-obituary?pid=96748256

Fig. 6.2 Virginia Dupies was profiled in *Dallas* as a woman working in oil (Thompson 1953)

The article also noted how excited Dupies was in funding scholarships. For instance, there was funding for women in geology at Southern Methodist University, as well as helping with scholarship in the field of petroleum engineering as well. She noted that the first scholarships were available to students in 1953–1954.

She was also active in getting programs in clubs like the Desk and Derrick Club that were serving the education interests of women in the oil field. The article framed her position helping women in the oil field: "She is also pleased with the outstanding programs served up during 1953 to the oil women who are members of this education minded club" (Kappa Alpha Theta Journal 1954, pg. 31; Thompson 1953, pg. 63). Dupies was quoted as saying that "Our purpose is to promote a clearer understanding of the industry in which we work, so that it will benefit our companies and our members, has been served by speeches and programs comprising some of the outstanding oil men in the Southwest including Mr. Frank M. Porter, president of the American Petroleum Institute, New York City, who has addressed the membership." But she noted that the meetings were not simply academic: "But it isn't all work and no play," she adds quickly. "We've had some wonderful field trips! Like the visit to the offshore drilling rig in Laguna Madre off the Coast near Corpus Christi, and the all-day trip through installations and refineries in East Texas. We learned a lot, but we had fun too" (Thompson 1953, pg. 63).

Dupies and her group participated in events that promoted oil, such as "Oil Progress Week." Promotions included spots on radio and television, as well as providing speakers for public groups and making "window promotion" about oil (Thompson 1953, pg. 63). Virginia Dupies also bragged that she was the only Texan in the group, especially her upbringing in Dallas. She was able to take her education at SMU, in which she graduated in 3.5 years of study, and transition to a business woman who was working in a company that was developing petroleum-based products. These products include many things like "…snake-bit kids and safety helmets to draw-works for million-dollar drilling rigs" (Thompson 1953, pg. 63). She was capable of working on many fronts: "And whether it is a matter of arranging for service, or sale of a derrick or drill collar, the vice president's secretary back in the Dallas headquarters of some 82 Continental stores and offices from Casper to Maracaibo can take the message or handle an order in driller's language if the 'boss' is out" (Thompson 1953, pg. 63).

Dupies contributed to work to sponsor the geology studies of female geologists. For instance, in *National Petroleum News* in 1953, she announced the scholarship of $200. The scholarship would be awarded to "…to one girl each year to study petroleum industry subjects at Southern Methodist University" (National Petroleum News 1953). The Desk and Derrick Club sought to be a positive civic group, as well as a space for industry insiders, and a place that opened up avenues of participation for women. The public relations and image arm of the Desk and Derrick club profiled women working in club as figures of aspiration and represented the "progressive" nature of the oil industry in regards to opportunities for women. There were other women like Dupies that sought to contribute to community but represented opportunities for women in the oil industry.

The oil industry, though publishing articles advocating the role of women in the home, continued to have ads and small articles appear in newspapers that changed the status quo of men dominating the labor market. For instance, in *The Cottonport Leader*, a small article ran in 1958 with the eye catching "The Welder is a Lady" and promoted the different type of jobs and access that women could have in the petroleum industry (The Cottonport Leader 1958). The story ran in a couple of newspapers from Louisiana to Missouri.[4] The article promoted that "The petroleum industry looks to woman for eight percent of its manpower." Boasting that there were around 132,000 women working in oil, they had broken into a predominately male-dominated field. The article used the same type of advertising and messaging as ads from World War II: "Currently there are more than 132,000 oil women, many of whom have crashed the "man's world," working as geologists, welders, landmen, engineers, paleontologists, service station attendants, research chemists" (The Cottonport Leader 1958). Then the article turned to how open the business was to women and how many opportunities in the petroleum industry were open to women. The career-focused woman should be attracted to the petroleum industry: "The progressive oil industry—with its more than 2,000 different skills and professions—is an ideal place for the career-minded woman whit the know-how ambition, and initiative" (The Cottonport Leader 1958). The petroleum industry was framed in the most glamorous terms. Also the headline of *Lady* frames the jobs and industry for women with sophistication and means.

Articles continued, however, in Texas and southwest-based newspapers. Many of the write-ups were reminiscent to stories about local do wells or national sports heroes. For instance, on June 7, 1953, in the Lubbock-based paper, *The Daily*

[4] IThe Dehli Dispatach *March 20, 1957*, *Albany Ledger* March 14, 1957, The Cottonport Leader, March 28, 1958, IThe Ville Paltte Gazette, February 21, 1957.

Ardmoreite ran an article about "Local Women in Oil" (Lang 1953). From industry product producers to explorers in oil praised the work of women. As much as the world of oil was a man's world, the messages in newspapers tried to use that perception as a marketing advantage, and oil companies actively tried to recruit women. The article opened with the enticement for female employees as the local community was really a great fit for women, "The oil industry and allied companies in Ardmore offer a great variety of employment to the women in this community" (Lang 1953). And all facts of oil companies want to recruit female workers: "The production, geological, and land departments of the many oil companies; oil field supply companies, oil field service companies, and other organizations closely connected to the petroleum industry all take advantage of the abilities and talents of feminine employees" (Lang 1953). The article promoted all of the different types of jobs that women could find for themselves in a thriving business: "There are [a] hist of niches in the oil business where women workers may find opportunity for a rewarding livehood and interesting occupation" (Lang 1953).

The article shows how a female worker took on a dynamic role with the company. The article profiled Quinne Cranford from Arkansas, who was working at The Texas Company, which would later become known as Texaco (Lang 1953). She started work at the Texas-based oil company after the war in 1951. She was working as a secretary for one of the head geologists John W. Mayes. While she was working at the Texas Company, she was serving in a secretarial capacity to seven geologists at a time. The job included administrative work, such as typing, and technical duties, such as map making. She had to design maps that included all the oil well work done in the Southern part of Oklahoma. And her boss John W. Mayes commented on how well Quinne did at that performance of her job, "Mr. Mayes tell us that his office has found Quinne outstanding in any job she performs and that 'she has exceptional organization and planning ability, and in the short span of time Mrs. Cranford has been with the Texas Company in Ardmore, she has reorganized and set up a completely new filling system'." (Lang 1953).

Crawford continued her interest, as well as her education, in geology and desired to study paleontology. She was a stellar student in high school, graduating from the locally based high school in Ardmore early at the age of 15, and pursued business education. She later studied art, as well as journalism, at the John Brown University, in Siloam Springs Arkansas, a city that boarders both Arkansas and Oklahoma. She has pursued correspondence courses in psychology, as well as typing, from schools in Oklahoma and Chicago (Lang 1953). Crawford had experience in journalism, working for the very same newspaper profiling her in the article (Lang 1953).

The article also praised her "career woman" status, as she was also raising two children at home. The writer listed a lot of Crawford's hobbies and accomplishments, such as decorating, reading, and outdoor recreation. She also started the Desk and Derrick Club in Ardmore. Much of the activities of Desk and Derrick were covered in *The Ardmoreite*. The article concludes with making the point that she is a representative, and aspirational figure, of women in the petroleum industry, much like June King, and many of the other women profiled in this book,

> While we present Mrs. Cranford for her activity in the petroleum industry, it is also appropriate that she be pointed out as presentative of the women in the business world who combined successful many roles—those of mother, homemaker and breadwinner and still is able to retain her induvial identity as an engaging and simulating personality. (Lang 1953)

Her article was composed by another major member of the Desk and Derrick club, Anita Lang. Cranford was active in her community and was participating in many civic organizations, which included Desk and Derrick (Lang 1953).

Lang would take on the role as promoter of the club and spelled out exactly what the purpose of the Desk and Derrick Club was in regards to its public perception. Anita Lang was a reported for the *The Daily Ardmoreite* and used her position there to report on her club and promote members like Quinne Cranford. She served as president of the organization in 1954 and put in the first part of the article the point of Desk and Derrick, which was "...to promote among the women employed in the petroleum and allied industries, through informative and educational programs, a clearer understanding of the industry which they serve, to the end that the enlightenment gained thereby may increase their interest and enlarge their scope of service" (Lang 1954). The activities, ad advertised, included lectures from " oil men," as well as trips and education, and interaction with newspapers through "support." It is likely that some of the newspaper articles about women in oil were encouraged or even written in the case of Lang, through plans by Desk and Derrick (Lang 1954). Speakers ranged from information about oil discovery to the guest appearance of All-American football players, discussions of oil created bazooka weapons during World War II, raising money for charity, and also mostly social events and dinners with oil companies like Haliburton (Lang 1954). Some activities included raising money for women seeking education in petroleum and geology, such as scholars money that was given in the amount of $300 dollars "...to a worthy young woman who desired to further her education in a field related to the petroleum industry" (Lang 1954). The scholarship was raised by the club hosting a "jean-lee style show[,]" which presumably was a fashion show associated with the Lee Jean company (Lang 1954). Some clubs, like the Long Beach Desk and Derrick Club, sponsored a beauty pageant. In profiling the Desk and Derrick regional meetings in *California Oil World*, a beauty pageant was described as "A petroleum version of the Miss Universe pageant entitled 'Miss Oil-Y-Verse' was the humorous entertainment furnished by the Long Beach Club following the dinner" (California Oil World 1959, pg. 20).

Towards the end of the lengthy and complete history of the Desk and Derrick Club of Ardmore, Lang framed the importance of the group in regards to history and the woman's place in the oil industry. Lang argued that Desk and Derrick was a change in the power of women but also their access to the industry.

> **Before the origin and growth** of Desk and Derrick, the feminine personnel of the oil industry resembled the man who stood too close to the forest to see the trees. We typed letters, took dictation, answered the phone, and went home—merely skimming the surface of all the tremendous meaning which this industry contains and overlooking the vital importance of what we were doing, trivial and unimportant as it may have seemed to us. (Lang 1954)

Lang also added that the true value of Desk and Derrick comes from the contribution and characteristics of women. Lang thought that the day to day work, which often gets unrecognized in the oil business and its related organizations like Desk and Derrick, comes from women, "The real worth of these clubs lies deep in the spirit and feeling they produce in those who become members, and which could not be produced were we not women" (Lang 1954). She also recognized that women were at work in the oil industry but not given sufficient recognition, "As has been said, women have not been an integral part for too long a time, and we have much yet to learn and many adjustments to make" (Lang 1954). She concludes with a more philosophical meaning to women in the oil industry and their participation in Desk and Derrick, "But these things we are realizing—that our jobs are not bound by four office walls, a desk, and a type-writer; that we are a vital part of something far greater than our individual spheres; that we, as women, complete with our foibles, do, can, and will continue to hold a most important position in the economic system under which we live" (Lang 1954).

Lang saw the Desk and Derrick Club as an organization that supported women and allowed to access a world of knowledge and participation in oil that was not formerly realized. It is likely that the Desk and the Derrick groups, ranging from Canada to Texas and Pennsylvania, continued the support and foundation for women entering into petroleum geology but also showed communities the contributions of women to the oil and gas industry.

Civic and professional groups blended in some cases in the post-war era. In today's society of the 2020s in the United States, we lack "third places" and the community organizations that made up the social landscape of the 1950s. The scholar Robert D. Putnam writes about the breakup of social groups and American life has focused on more individualistic pursuits in his book *Bowling Alone: The Collapse and Revival of American Community* (Putnam 2000).

6.3 Petroleum Peggy

During the 1950s, a weekly column ran in newspapers entitled "Petroleum Peggy" (see Fig. 6.3). Peggy was the pen name of Ann Taggart. However, the column was started at the behest of oil executives and newspapers editors in Kansas and Nebraska during 1949. The aim was "…to publish a series of readable stories concerning the wonderful way in which petroleum is helping to make the life of the average woman easier, brighter, and more interesting" (The Healdton Herald 1954a). Taggart was then appointed to write in the name of "Petroleum Peggy" as she was working as the executive assistant to the "General Secretary" of Mid-Continental Oil & Gas Association (The Healdton Herald 1954a). The article made note that she was either in charge of writers of the columns or the writer of the column herself. The 1954 article shared the news that as of August 1, the column was ending. But the column had been very popular, according to *The Healdton Herald*. Petroleum Peggy received a lot of letters from readers and editors and "'Petroleum Peggy' has in a

Fig. 6.3 Column image for the Petroleum Peggy Column (The Frederick Press 1956)

little less than five years become a very real person" (The Healdton Herald 1954a). The column then started a publication range of 48 states and now will become a more news-based column of the regions of the oil Continental Oil & Gas Association called "Notes from Petroleum Peggy" (The Healdton Herald 1954a). The column would now be under male editorship. Other columns would be offered called "Petroleum Pete Says—". Petroleum Pete columns would become more regularly appearing in newspapers (The Healdton Herald 1954a).

Petroleum Peggy commented about many things, from fashion to cars, but often tied her column into the virtues of the petroleum industry. In February of 1954, only a mere couple of months before the column was discontinued, there were three columns that appeared in *The Healdton Herald* (The Healdton Herald 1954b). One column discussed a woman's knowledge of the car, as her knowl4edge had been increasing. The woman of the house was often instrumental in selecting the car, and the column reminded the reader with pride that it ran on gasoline. The column then instructed the female driver to get the most mileage out of her car over six points. Petroleum Peggy then reminded the female reader that cars led to freedom: "Enjoy your car! Nowhere else in all the world can people travel as freely, conveniently, and economically as in these United States" (The Healdton Herald 1954b).

Another column, in the same paper on the same day, promoted the importance of the petroleum industry to women's fashion. Peggy noted the importance of petroleum workers to the unpredictable world of fashion: "It takes more than a crystal ball to predict fashions Millady will wear! An army of people have been trying to read her mind, however, to determine what will appeal to her in the spring of '54. Want to hear some of their best guesses?" (The Healdton Herald 1954b) The "oil wells" had replaced "silk worms" and now Nylons were made from oil by products like cyclohexane (The Healdton Herald 1954b). The stocking came in many colors and were increasingly more sheer. And Peggy reminded the reader that the oil company was working specifically for her: "And oil people maintain the close watch necessary to manufacture cyclohexane. All for YOU—Milady—and your new spring fashions" (The Healdton Herald 1954b).

Other columns in 1952 promoted the idea that Georg Washington was the firs "oil man" as he saw the use of lubricants in the eighteenth century (Peggy, Petroleum 1952). Peggy provided another column that promoted the importance of toys in the development of children in *The Wilson-Post-Democrat* on December 25, 1952 (Wilson Post-Democrat 1952). Children, at a young age, learn a lot their toys and "playthings." Toys often reflect the type of work that the children might pursue later in life, "I always think of children's toys as being playthings, but some of the them look more like work-0things to me. They educate Junior while they entertain him, and they allow him to imitate his elders in helpful ways" (Wilson Post-Democrat 1952). Peggy noted that while tots learned about bulldozers and toy cars, little girls learned about household management with kitchen sets

> Little girls keep house "just like mother" with a pretend kitchen set that includes refrigerator, range, and sink with such fascinating gadgets as a dish drying rack, ice-cube trays, and a pressure cooker. Many of these and other equally enchanting miniatures are made of plastic, a versatile material that comes from petroleum. (Wilson Post-Democrat 1952)

Peggy reminds the reader that all children need toys to learn and those toys from plastics from the petroleum industry. But she ends the column in an interesting way, cautioning parents that the toys that they select for their children must be done carefully, as they need to fit their needs and growth. The ending caution could also be read as a word of thought about putting boys and girls onto specific labor tracks that might not fit the child.

> The problem here is that parents often consider their own tastes instead of their child's needs. They have a disturbing tendency to buy an elaborate doll dress in satin and lace for a three-year-old who would love a simple, cuddy toy of washable plastic. Or, they get a convertible, slightly less than full-size, and complete with gasoline motor, for Tiny Tim, age five, instead of for Joe Junior, age ten, who years to be a mechanic. (Wilson Post-Democrat 1952)

Peggy advised the reader to simply observe their children's play for ten minutes a day and see what types of activities they move to and then get toys that would complement their interests.

Peggy promoted the use of cosmetics as well. Other columns promoted the use of helicopters and the creation of plastic miniatures. Petroleum Peggy often had larger newspaper spaces entitled "A Woman's World and Oil," where she was able to promote the variety of product colors and paints to women thanks to the petroleum industry (Peggy 1959; The Frederick Press 1956). The oil industry was crucial in promoting a proper home environment as the choice of colors was pleasing, according to Peggy, and she asked her reader: "What color do you feel today? The question may sound like nonsense, but it isn't. Modern science is coming closer and closer to the ancient beliefs in the power of color. The colors in your home can make you cheerful, or relaxed, or depressed, or even argumentative!" (Peggy 1956)

Petroleum Peggy also had political learnings, as she did not agree with communism. In a column that discussed jobs for women, she also tilted at communism.

While Peggy was excited about jobs for women, she was surprised that women were working in iron foundries: "We're always glad to hear of new job opportunities for women, but here are some that rocked us back on our heels. Women are being employed in iron foundries in Poland at open-hearth furnaces—jobs that are among the most dangerous and strenuous in heavy industry" (The Healdton Herald 1954c). Working in an iron foundry was physically taxing for women and required a lot of resilience, and the heat was intolerable. Poland was, as Peggy pointed out, "...a captive country under Communist rule," which other "professions" had started to employ women, such as mining, garbage hauling, and masonry (The Healdton Herald 1954c).

Women working in these countries did not have access to the technologies of the West, which Peggy included in a tongue-in-cheek way as "decadent West," as she recounted the hard labor required of women "...in these captive Communist countries..." (The Healdton Herald 1954c). Women still had to work in the kitchen in countries like Poland, again free of modern technologies, and Peggy added that husbands often do not help as they feel it is not their place to be in the kitchen. And the stakes, according to Peggy, are quite high for women's domestic work, "If a woman's efforts do not meet Communist approval, she may be sentenced to a force labor camp for as long as 15 years" (The Healdton Herald 1954c). The column then differentiates America to Communism, as there was private enterprise and "great industries" such as petroleum. It was, according to Peggy, to private ownership and profitable industries that there was a good life in American and lots of "abundance" (The Healdton Herald 1954c).

Women were active in the physical offices of oil companies. Office and secretarial work allowed women a place in directing some of the focus of the petroleum industry. These women often aimed the industry at serving the needs of customers like them. Creative opportunities continued for women in office work as they became increasingly involved in building office culture.

6.4 The Petroleum Engineer: Women at Work

Publications for the oil industry promoted the work of women throughout the Second War and into the post-war period. Publications included profiles on women, and there were even contest where women commented on their roles in office work. In *The Calgary Herald* in 1953, the results of the Desk and Derrick Essay contest were published (Calgary Herald 1953a). The winner was Agnes E. Bracher (later Barrett) who studied business at the Texas Women's College. Bracher worked for Humble Oil, which later became Exxon (Office of Alumni Engagement 2022). She was recruited by Humble Oil to work in petroleum during the war years. The name of her essay is "Why My Boss Should Send me to the Oil Show" (Calgary Herald 1953b).

The *Monroe Morning World* ran some of the memorable lines of Bracher's essay on May 3, 1953 (Monroe Morning World 1953). The essay was a description of the perceptions of experiences of the women in the offices of oil companies. Bracher described her environment, "'You see us everywhere. The trim little blonde who brings your mail during the day is the Hall Girl. The lady in the white uniform that gives you the red pill and swabs your throat is the Industrial Nurse.[']" (Monroe Morning World 1953). She also pointed out that there were women working science jobs at the office, and those jobs utilized microscope, "Peering through her microscope and juggling test tubes up the next floor is the chemist." She continued to describe the other rhythms of the office, "Beating out a tattoo on her typewriter in the next office is the stenographer" (Monroe Morning World 1953).

Then Bracher turns to the women drawing out the locations of the oil wells and the geologists, "Down the hall working on a map to select the spot to 'shoot the well' is a lady geologist" (Monroe Morning World 1953). And the administrators of the office, "And the efficient young woman who makes your appointments and keeps your personnel records is your secretary.[']" (Monroe Morning World 1953). Bracher ends her essay by stating that all the value of science and progress in the oil industry will be observed and studied by her and then brought back to the company to improve the office. She argued to her boss at Humble Oil that, "I would actually see portrayed nearly a century of progress in the oil industry! Boss let some of that Tulsa oil lubricate this little 'cog' in our company. I'll come back geared to do a better jo for you and for Humble" (Monroe Morning World 1953).

The initial announcement for the essay content also ran with another column by Henry D. Ralph, a reporter at *The Oil& Gas Journal* that discussed the changing nature of the oil show and how much more seriously "oil women" are considered and a part of a professional show. He drew a sharp contest to the earlier oil shows, where women participated in types of beauty pageants and rode around on "decorated conveyances," or floats representing different oil producing regions trying to win an overall beauty crown for the oil show (Ralph 1953). Ralph ended the article by making the point that now women were a major part of the petroleum industry, "All of that sort of stuff is out, now. The oil show itself is an industrial exhibition for technical operating men who come to look at mechanical equipment and study performance figures" (Ralph 1953). He then emphasized that an oil show is now a serious professional meeting that women are a part of, "And the oil women present will not be there because of their good looks—though they'll have that, of course—because they are now a real part of the operations of the oil industry" (Ralph 1953).

He also reported that Desk and Derrick got the idea of giving the award of "Oil Woman of the Year" (Ralph 1953). She would be given the award for her contributions to the oil industry and was implied not her beauty, as the title of the column was called "Brains vs. Beauty," as Ralph tried to juxtapose, or promote the perception, of the changes going on in the industry for women (Ralph 1953).

Women's career prospects were going through more general changes in 1956. *The Spokane Chronicle* ran a piece, written by a female writer, which discussed the new working woman of 1956 (Toomey 1955). Statistics about the woman of 1956, though these statistics are forecasted in 1955, tried to explain the working life and upcoming experiences of women. Toomey wrote that because of these statistics, there is a lot of knowledge about the experiences of women: "The woman of 1956 will be surrounded by less mystery than any female in history" (Toomey 1955). She continued that, "Never have so many statistics been stacked up in a single year on supposedly unpredictable female as we accumulated in 1955" (Toomey 1955). The only thing, Toomey pointed out, that statisticians have yet to examine is the "moods" of women.

Because of so many changes in women's lives, the report noted that women were experiencing some stress. Toomey, reporting on government studies, summarized that, "Women are jittery, the study group reported, and likely to remain jittery until they define their role in man's life and quit trying to be career women, mothers, civic leaders and glamour girls all at once" (Toomey 1955). Toomey was likely commenting on the pressure that women experienced having to be both accomplished at work and beautiful, as there are many examples reflecting those pressures in this chapter.

Women's lives were also very busy and statisticians could forecast their major life changes. Toomey noted that statisticians had found that though, "A woman may not have the foggiest notion what she is going to be doing in the next 12 months, yet the facts show that: If she is 20 years old she is likely to be a bride in 1956. Her bridegroom will be three years her senior, according to the law of averages" (Toomey 1955). She will likely be a working woman, even if she is married, and will continue to work through her marriages. It is likely that she will have a child in the first three years of marriage. Her number of children will often rise to two children as it was estimated that it was four times as likely, compared to her mother, that she will live in a four-person household. It was also likely that she will live to her seventieth year. Her work will likely take place in an office and likely being some type of clerical job, as clerical work was dominated by women. Her salary will be around one thousand dollars a year. This, compared to her husband's earnings, is less, but because she is working, "….her salary will make it possible for them to furnish their first home in new contemporary designed furnishings" (Toomey 1955).

Her office will be made up of mostly female employees. Toomey cited research that found that the office would be made up of three women for every male employee. Women would take leadership roles in the office, especially compared to their historic peers: "She has more chance[s] of dominating her husband than grandma had of dominating grandpa. Nine out of 10 husbands reported they helped working wives with the house work" (Toomey 1955). The working woman will purchase clothes, around four new dresses a year. In a paradox, researchers found that the working woman of 1956 will be "healthier" compared to husband but get sick more often than her husband.

More women were working or returning to work over the age of 40, almost at twice the rate of current working women. Women over the age of 40, Toomey extrapolated from the research, would have trouble finding a job. Some jobs in offices and clerical work ran adds that wanted employees who were age 35 and under that age. But other companies are starting to make exceptions. In terms of family finances, one in every ten women will be the major bread winner for the family. Some of these women are carrying financial responsibilities, as it was estimated that one in every four women over the age of 50 would be widowed. In this chapter, and others preceding it, physical characters of women are often always paired with discussions of work, so Toomey included information from statisticians that "Her average height is five feet three inches and her other vital measurement are 35-29-38 (that's what the government says)." (Toomey 1955).

Women working in offices were also going through profound professional, social, and societal changes. The oil industry and petroleum geology both reflected those changes, and the women working in all aspects of oil and geology work experienced those changes at their jobs and personal lives. The 1950s also showed a transition point in both the experiences of female employees and the opening of work to women in the 1950s. Women were also pressured to do it all, be both glamorous and professionally competent. The next chapter will further explore women working petroleum geology both in the office and in the field. It will examine how women transition from the pressures of glamor to making their own spaces and culture in the office and will examine how the oil industry was transition to the new women of 1956. The oil woman was changing with the times as well.

References

(1952) Petroleum peggy says. The Wilson Post-Democrat. 25 December
(1953) About oil people. National Petroleum News 45(15):104
(1953a) Oil industry's 'first lady' to be chosen. Calgary Herald. 11 February
(1953b) Houston secretary wins D. and D. essay contest. Calgary Herald. 15 May
(1953) Award given for essay in Desk and Derrick contest. Monroe Morning World. 3 May
(1954) Virginia dupies carves career in oil industry. Kappa Alpha Theta Journal 68(3):31–32. This is a reprint of the article that appeared in Dallas magazine. See the citation Thompson, Claribel
(1954a) Petroleum peggy says. The Healdton Herald. 19 August
(1954b) Petroleum peggy says. The Healdton Herald. 11 February
(1954c) Petroleum peggy says. The Healdton Herald. 17 June
(1955) First oil educational institute is held at college in Tonkawa. The Ponca City News. 12 October
(1958) The welder is a lady. The Cottonport Leader. 28 March
(1959) Oil industry in Canada and U. S. A. is synonymous. Leader-Post. 28 March
(1959) Desk and Derrick clubs hold regional meeting. California Oil World 52(12):20
(1961) Industry woos students with science skills: big firms scan education facilities in locating plants. Toledo Blade. 21 May
(1962) A feminine wheel in oil. Toledo Blade. 7 January
American Petroleum Institute (1951) 8, 179 new oil inventions patented by scientists in five years. Life Magazine. 10 September, p. 143
American Petroleum Institute (1953) Man Hunt: Oil Companies Complete for the Class of '53. Life Magazine 9 March, p. 150
Bucher CA (1960) Looking ahead to college. Seventeen Magazine 19(6):134–135
Dawson S (1961) Business mirror: skilled hands in demand. Spokane Daily Chronicle. 22 May
Doan C (1961) Beauty consultant gives sage advice. The Leader-Post. 21 April
Lang A (1953) Local women in oil. The Daily Ardmoreite. 7 June
Lang A (1954) Rambling reporter. The Daily Ardmoreite. 29 August
Office of Alumni Engagement (2022) Agnes E. (Bracher) Barrett '44. Alumni Spotlight. https://twu.edu/alumni/alumni-spotlight/agnes-barrett/
Peggy P (1952) George Washington saw future for oil. The Ponca City News. 14 October

Peggy P (1956) A woman's world and oil. The Frederick Press. 9 August
Peggy P (1959) A woman's world and oil. Eastern Montana Clarion. 29 February
Matelski EM (2017) Reducing bodies: mass culture and the female figure in postwar America. Routledge, New York
Stanford-McIntyre S (2022) Desk and derrick: the women's petroleum industry club that envisioned oil's technocratic future. Labor 19(4):6–26
Putnam RD (2000) Bowling alone: the collapse and revival of American community. Simon & Schuster, New York
Ralph HD (1953) Journally speaking. Oil Gas J 51(36):53
Ravner PC (1953) "Older" women as office workers: training programs in four cities, facts on "older" women in relation to office work. Women's Bureau, Washington, DC
Standard Oil Company (1950) We send money to 40,000 women. The Telegraph-Herald. 24 September
Toomey E (1955) Woman's view: satistics dispel mystery about the woman of 56. Spokane Chronicle. 30 December
Thompson C (1953) Women in business: Virginia dupies. Dallas 32(10):63
Wdowik M (2017) The long, strange history of dieting fads. The Conversation. 6 November

Chapter 7
"Mobilogue": Women, the Office, and the Oil Industry: Women's Changing Place in the Office

7.1 Promoting Women in the Oil Industry

After World War II, women were pressured to be both glamorous and working women, and petroleum geology was no different. Trade periodicals promoted the working and glamorous women, and many of the women embraced that identity (War living: aprons 1943). However, in the postwar period, women would also push against aspects of identity in the workplace, such as in the case of office fashion and attire. Women in the postwar, spurred by advances in the women's movement, will continue to forge their own identity in the workplace, working with the oil industry when necessary to continue to assert themselves in the workplaces of the office and the field.

Trade magazines highlighted women working in the postwar oil industry, and both women and industry used each other to craft the image of the glamorous working woman in geology. *The Petroleum Engineer* ran a special section entitled "Women at Work" where they highlighted women in the oil industry (The Petroleum Engineer 1955, pg. E-11). These features often contained short biographical information about the type of work that women were performing in geology, or geology linked tasks, and they were highly technical (see promotional image in Fig. 7.1).

One notable example was Olivia Whitehurst, who was profiled in the 1955 first issue of *The Petroleum Engineer*. Whitehurst, originally from Shreveport, Louisiana, worked as a "geological draftsmen," and the publication remarked that she "…is one of the few women geological draftsmen ion the country" (The Petroleum Engineer 1955, pg. E-11). She transitioned from a career teaching high school history, as she had a longstanding talent in art, and her brother, already working at the Texas Eastern Transmission Corporation, encouraged her to learn about geology and perfect her artistic talents. The World War changed her career trajectory, and her husband joined the United States War effort to serve in the Coast Guard. Whitehurst used the time to train in engineering and drawing, as she took night courses. She

E. A. Driggers, *Glamour and Geology*,
https://doi.org/10.1007/978-3-031-64525-9_7

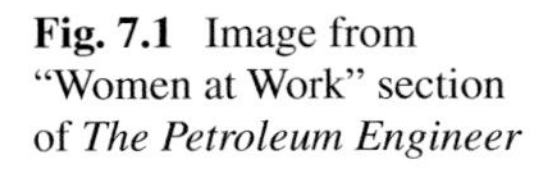
Fig. 7.1 Image from "Women at Work" section of *The Petroleum Engineer*

started as a draftsman for an electrical engineering business. A draftsman is someone who makes technical and engineering-based drawings (Shelmire 1919).

After the war, she graduated with a degree in geology from Centary College and also studied at the Southwestern Institute of Art. She became a geological draftsman shortly after her time in the electrical engineering firm. Her job mostly focused on making maps of the location of oil fields and she was very proficient at her job: "She is quite at home making maps of various oil and gas fields and prospective productive areas, laying out stratigraphic sections, and keeping regional maps up to date with the latest well and lease data" (The Petroleum Engineer 1955, pg. E-11).

The publication, in keeping with the style of the War period, also profiled her personal characteristics as a woman (see Fig. 7.2) and emphasized how her hobby both supported herself and made for a proper family situation. The profile ended with a write up of Whitehouse's contributions to her family and her character as a mother:

> Olivia spends most of her spare time with her favorite avocation—nine year-old Sherry Whitehurst. She occasionally finds time for her hobby of photograph painting, and recently she completed a project that she considers the best-loved drawing project of her life—that of designing the dream home where she and her family now live. (The Petroleum Engineer 1955, pg. E-11)

Women like Whitehurst benefited from employment, experiences, and retraining opportunities during the World War that led to postwar prosperity for white, middle class, women. Like many women in this chapter, these women, were discussed and promoted in trade publications and other news sources as having a female duality of the highly technical while also exhibiting the virtues of womanhood that countries like the United States embraced during the middle of the twentieth century.

7.2 Women, the Science of Oil, Research, and the Office

The Petroleum Engineer also focused its "Women at Work" series on women excelling in the office and at business (women who were profiled appear in Fig. 7.3). In the 1954 volume, women working in both science and the office were promoted to

Fig. 7.2 Whitehouse pictured in *The Petroleum Engineer* (The Petroleum Engineer 1955, pg. E-11)

Fig. 7.3 "Women at Work" featured working women (The Petroleum Engineer 1954a, pg. E-36–E-37)

the reader and the short article explained how each woman made major contributions to the oil business. Mary Alexander was described as having “The Magic Touch,” when it came to the chemistry of petroleum, and having a job that was “…one of the fairy-tale kind.” She was continually described as magical with her abilities, “It took magic to create it and she has developed a magical touch keeping it” (The Petroleum Engineer 1954a, pg. E-36–E-37). Mary Alexander was “…in petrochemical research” (The Petroleum Engineer 1954a, pg. E-36–E-37). Like other women discussed in this chapter, she started out studying journalism. Newspapers and magazines were a very open field for women, and major figures during the postwar era got their start working for magazines and newspaper, such as Jacqueline Bouvier, who would later become First Lady of the United States (Anthony 2023; Whitt 2008).

Alexander had to take math and science as general education requirements and found that she really liked them, and then she changed her career trajectory, and found that trying to take the general education requirements quickly exposed her to love math and science, “…to get them out of the way so I could get on to more interesting subjects” (The Petroleum Engineer 1954a, pg. E-36–E-37). The article went on to record that, “Body was more surprised than she when she discovered that she loved chemistry” (The Petroleum Engineer 1954a, pg. E-36–E-37). Then she went to study at Texas Tech and then at the University of Iowa, when she then landed a job working with the director of research at Universal Oil Products Company.

Alexander then was successful, especially during the war years. The article highlighted how she perceived her position and herself as fortunate, “That was my lucky day,” she says. “My work has been fascinating, my associations inspiring, and the oil industry intriguing” (The Petroleum Engineer 1954a, pg. E-36–E-37). She went on to further highlight how lucky she was about having the job, and the publication continued to amplify those ideas, “I could not begin to describe everything the job includes. We never know from day to day what will come next. Working with Dr. Egloff is stimulating and I consider it a unique privilege” (The Petroleum Engineer 1954a, pg. E-36–E-37). She also commented on the editorial and writing parts of her job, which she embraced because she enjoyed living in the city, “…I have not gone so far afield from my original plans. My work includes considerable writing and editing and through petrochemicals, my interest in textile chemistry has been renewed” (The Petroleum Engineer 1954a, pg. E-36–E-37).

In the 1955 edition of *The Petroleum Engineer*, the idea of the male mentor as crucial to the development of women in the petroleum industry was promoted in the issue (The Petroleum Engineer 1955). Dorothy Harbison worked in the oil industry in Houston. She was hired by Dr. John D. Todd during the war in January of 1945. Even though Harbison graduated valedictorian at her high school, Todd explained when referring to Harbison to *The Petroleum Engineer* in a paternalistic tone, “took her to raise,” but the magazine pointed out that both Todd and Harbison were content with their situation (The Petroleum Engineer 1955). Harbison in 955 worked for Todd as his secretary but did much more than simple work: “—among them setting up all lease files, approving payments of drafts, sending instruments for record, paying rentals, ordering title certificates and doing preliminary curative work before the certificates are turned over to an attorney" (The Petroleum Engineer 1955). She was managing oil land records, royalties, and other complicated records for Todd, even having properties across several states. She managed many aspects of Todd's business, as she was working to help "...in the management of Cambe [Cambe Log Library], interviews all new personnel, makes recommendations for promotions, supervise bookkeeping department, manages the geological book department" (The Petroleum Engineer 1955). When the reader surveys this profile, they likely question what, if anything, did Todd do at the company. The profile ended with a profile of the woman's personal life, as they often included in these profiles, and emphasized her relationship and femininity. She was pursuing an undergraduate degree, studying oil law, and managing her garden issue (The Petroleum Engineer 1955).

In 1954, some women were profiled as being involved with oil exploration and uranium discovery. Frances Oldham (Fig. 7.4) was described as being "...living proof that a woman can handle about any job given to her" (The Petroleum Engineer 1954b).

Her boss had much confidence in her abilities (see Fig. 7.4). She held several royalties and owned several acres of land, and was one of the richest women in Wyoming who was working in oil. Her business judgement was continually praised by her boss and co-workers. She also speculated for the business and her boss was confident in her judgement,

Fig. 7.4 Francis Oldham pictured in *The Petroleum Engineer* (The Petroleum Engineer 1954b)

> She has authority to buy and sell land when her employer is out of the office and has entered into a number of separate business ventures in his absence. She acquired a larger number of uranium and vanadium leases on her own initiative and two years later Mr. Anderson became one of the owners of the American Uranium Company. (The Petroleum Engineer 1954b)

The archetype of the genius employee and absent boss emerged through the pages of the *The Petroleum Engineer*. Other issues of *Petroleum Engineer* advertised in this focus section the multitude of ways in which women could be involved with the oil industry, such as owning their own companies, working as oil attorney, or helping to run oil trucks (The Petroleum Engineer 1953).

When Alice Brady was profiled, the common trope of presenting the false dichotomy of beauty and highly technical jobs as surprising the reader was utilized in the 1953 profile. Brady was described as a leader in making sure that all transportation of oil and fuel went successfully to market, as she managed many, if not all of, the facts of getting oils from the fields, to the refineries, and then to their destinations (The Petroleum Engineer 1953). *The Petroleum Engineer* reminded its readers of the anomaly of a beautiful woman doing a highly technical job, "You just don't imagine pretty girls like Alice would want to bother about where a tank car of fuel oil was headed but she thinks it is so exciting that she has been with Ben Franklin Refining Company 10 Years, beginning as a stenographer, and she studies all the time to do her work better" (The Petroleum Engineer 1953). She was also involved with Desk and Derrick Club in its leadership.

The article that ran right beside the "Women at Work" section promoted the surprise of the highly technical male engineer, that simply was not the stereotype of drinking and womanizing (The Petroleum Engineer 1953). The article, which discussed the first Petroleum Engineering School at Oklahoma University, tried to dismiss the hyper masculine assumption about a petroleum engineer: "A petroleum engineer, it is thought, has always been sort of a jack of all trades, a poor boy's geologist, surveyor, construction boss, electrician, chemists, lawyer, and judge of fine whiskey and beautiful women. There must be some reason for this diversified educational background" (The Petroleum Engineer 1953).

The difference between the two stereotypes of masculinity and feminity, which apparently were so commonly understood that *The Petroleum Engineer* discussed them the stereotypes as proper social values for men and women. The male petroleum engineer was sophisticated and boot-strapping, a brand of Hugh Hefner, which tried to define the 1950s and 1960s masculinity through good taste and womanizing, or a type of debonair sophistication, that most historians could cast as a type of toxic masculinity today (Osgerby 2001; Harrington 2020). The phrasing in *The Petroleum Engineer* implied that beautiful women were something to be collected, like whiskey, or even owned. The troupe of the new beautiful woman scientist, who both works in a high technical field because she choses too, and not because of her lack of interest from the male gaze, both entrapped women into new and more restrictive societal expectations. Women working in science now had to have a hyper fixation on their appearance, but also experienced pressure on their roles in the office and in petroleum geology. Women now had to be the most exceptional worker, like Alice

Brady, who worked constantly to improve her job performance (The Petroleum Engineer 1953). I would argue that the later female stereotype of the woman working in the oil industry persisted and became even more pressure came with the new role of the high achieving "oil woman."

7.3 The "Oil Woman"

Winnie Johnson was described as an "oil woman." Her advent for adventure was praised, along with her femininity, popularity, and marriage status, "She has made most of the breaks in her life and her diligence and vision have helped her to walk with confidence through every door opened to her. She is an oil woman, a cattle rancher, and an aviatrix. She has studied business, interior decorating, the oil industry, and Spanish. She has won a silver medal for scholarship and a diamond studded pin for personality. She also proudly wears a wedding ring" (The Petroleum Engineer 1954a). She was described as the "Personality girl" pictured in Fig. 7.3. The description likely would have included the phrase "accomplished" had she been born in the nineteenth century. Currently, she was working "right-hand-man" to an oil president, and also managing the accounting books of the Baker and Yalor Drilling Company, as well as managing ranches for her boss. Her previous career activities also included serving as a stewardess on American Airlines after graduation. She happened to meet her future boss while visiting her sister. Her boss is often out of the office and she was there to lead the office,

> Her new employer was out of the office much of the time and just any new girl might have found the job a marvelous opportunity to loaf, treat herself to long lunch hours, and think about her social life. Not Winnie. She armed herself with every book and magazine she could find on the oil industry, asked questions of everybody she thought might know any answers. Then she went out to visit two rigs in operation and "Learned more in three days than weeks of book study could have taught men." (The Petroleum Engineer 1954a)

Johnson was also a member of the Desk and Derrick Club and was extremely active in the chapter in Amarillo.

The term "oil woman" started to rise in popularity after World War II. Two searches on Google's N-gram shows the popularity of the term and its increases. This is especially true in the rise of the 1950s. The Google N-Gram analyzes (Figs. 7.5 and 7.6) all of the materials scanned into google books, such as books and periodicals, and tracks how popular a world becomes.

Women like Johnson would become incredibly important in promoting the oil industry, continuing and expanding the image of the woman in the oil industry and petroleum geology (The Petroleum Engineer 1954a). The next chapter will focus on women working in the office and promoting oil, and working in organizations like Desk and Derrick. The concept of the "oil woman" will continue to rise in popularity as well.

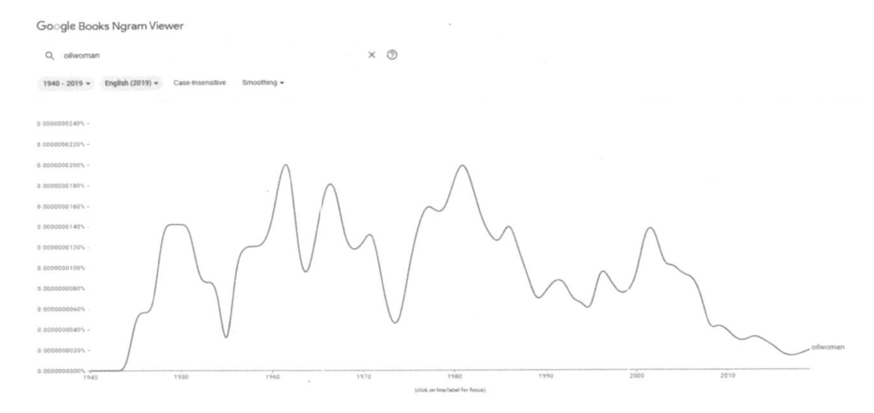

Fig. 7.5 Ngram results from 1940 into the 2010s

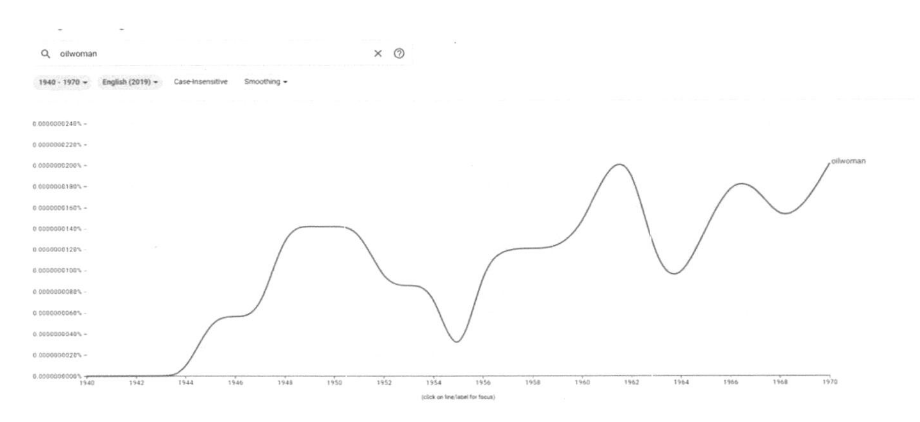

Fig. 7.6 NGram results when they are focused from 1940 to 1970

7.4 Mobile and the Recruiting of Women

In 1945, Mobile ran a glitzy ad in *Glamour* magazine entitled “Women In Oil Jobs.” The reprint contained a summary of the ad, “Glamour magazine carried this two page feature in July 1945 as one of its regular stories on fields of work. Pictures and text describe unusual jobs women are doing in the oil industry” (Glamour 1945). The profile was very typical of other flyers, media campaigns, and articles featured in this book. June King was again, featured in the profile entitled “Strike Oil” (Glamour 1945).

Glamour was a magazine that had a mostly female audience. The ad was targeted not to advertise the viability of jobs, but that these jobs in the oil industry would contribute to the applicant’s personality and be much more than a job. The ad advertised that a stenographer could work for any boss and take notation, but working in the oil industry is much more of a calling. The advertisement had the tone of an armed forces recruiting ad,

> A stenographer can take dictation whenever there's a boss with a letter on his mind, a booker can make a trail balance so long as she has a set of figures, a receptionist can smile politely behind anybody's front desk—*or* she—and every other specialist—can look at her job in a broader, brighter light. (Glamour 1945)

The work was more than work, but a major field of work that had a positive future for women. Women had talents that the oil industry would utilize in order to make progress. The article defined both technical and clerical positions as having a cache of prestige, asking the reader to visualize herself in a job within the oil industry, "Can see it, not merely in terms of *what* she does, but *where* she does it. Can choose deliberately to use her special talent in a field-with-a-a-future" (Glamour 1945). The women that appeared in the pages of the ad have done just this, brought their talents to a job with a promising future, and the oil industry was a place that had expanding boundaries where women could work. The propagandistic nature of this claim was exaggerated, as it took court orders in the 1970s to truly expand the industry to women, but the opportunities in the oil industry, compared to other industries, including academia, seemed open, especially during the late 1940s and early 1950s.

The ad enticed women to find their place in the oil industry and become satisfied with their work, even in jobs that are diverse and use different types of skill sets: "The young women on.

These pages, for instance (who skills range from medicine to metallurgy) have all found their places within the wide, elastic boundaries of the oil industry. Oil is colossally Big Business in peace or war, and research continues to turn up new by-products yearly" (Glamour 1945). Women would always have job security, the job implied, because of the needs of oil, no matter the security situation of the nation.

Jobs in the oil industry were expanding for women and women could make a comfortable living. The ad spoke to an optimism of employing women in the oil industry, "Consequently, jobs in Oil multiply almost as fast as rabbits in the field...pay off in social security programs, promotions, frequent wage raises. The girls here are on-the-job throughout the country for Standard Oil Company (N. J.) or one of its affiliates" (Glamour 1945). Women were invited "To Try your own talents in the oil industry[.]" (Glamour 1945). Profiles followed of the diverse jobs needed in the oil industry from reporter, laboratory tech, nursery school teacher, translator, mapmaker, photographer, geologists, and other jobs related to laboratories and research. Each woman was profiled to give the reader a sampling of what type of work she could apply her talents to in the oil industry (Glamour 1945). The profiles read much like the "Women at Work" profiles that appeared in *The Petroleum Engineer* in the 1950s.

In discussing the work of a geologist, June King was featured. The profile discussed her working in both the lab and oil field, and that she was the single female geologist currently employed by her oil company. Unlike the write up in the *Weekly Standard,* this profile emphasized that she did most of her work in the laboratory, but did occasional field work, "Like most women of the profession, she does much of her work in the laboratory, but she enjoys occasional field trips, looks forward to some day discovering a well" (Glamour 1945). The description in the *Glamour* magazine lacks the pin-up culture and aggressive nature of her performing her job. This might indicate that the *Weekly Standard* was targeted at general readership, or even the male gaze, while this ad, likely composed, or at the very least contributed by women like the reporters and photographers mentioned in the ad, was designed for recruiting woman.

The male gaze often frames photographs, stories, clothing, and culture. This idea first originated from film critiques. A good statement she recorded from another scholar of the idea of "the male gaze" comes from Lauren Michele Jackson's piece about her experiences seeing films that were made "...in a manner hostile to my flourishing as a woman[,]" where images on film serve to "....reinforce patriarchal fantasy" (Jackson 2023). Therefore, images, poses, and the positioning of women in profiles of female geologists seems to serve heterosexual male ideas of attractiveness and sexual desire. There is a lot of scholarly debate if pin-up photographs are empowering to women or are somewhat pornographic (Buszek 2006; Levy 2006). Aspects of pin-up culture is associated with both the male and female gaze, it depends on the photograph, photographer, the producer of the photography, and the intended viewer. Perhaps oil companies recruiting female photographers wanted to produce images of women working in geology that would be of interest to women and help with their recruiting goals. Other readers of this book might see the oil companies as employing women to continue to produce images that were of interest to men.

Photographer Esther Bubley got her start taking photographs for the petroleum industry, eventually transition to working for *Life*. The 1945 profile of Bubley included that she was mostly responsible for photographing the war industry (Glamour 1945). She was trained in paining and sculpting, as well as photography. She also had experience working for *Vogue Studios* and OWI. She was responsible for the images in the *Glamour* magazine profile (Glamour 1945).

Profiles also included non-technical positions that supported workers in the petroleum industry, such as that of Cornella Stryker. Stryker worked in the nursery school, and the ad advertised a certain level of glamour to her job, such as working in "sunny Aruba" (Glamour 1945). She was university trained and had four years of teaching experience in California prior to her taking her current job. The oil industry also hired translators like Janet E. Jackson. She was reading and translating oil related to oil in both French and German (Glamour 1945). Jackson also organizes research papers and serves as an organizer for major meetings of the petroleum industry. She brought her undergraduate education in chemistry to aid in her transitions of petroleum papers. Catherine Munn served as a reporter and promoted the oil industry through publications like the *Esso Dealer*. Munn wrote stories about gas station attendants and traveled around the East to interview workers in the petroleum industry. Researcher Margaret Hankins was a statistician, who also assisted in the preparation of articles about petroleum and the petroleum industry. Prior to coming to Standard Oil, she had worked in research for *Time Magazine* (Glamour 1945).

Researchers Fredricka Lofberg and Edith Keiper worked in laboratories. Keiper was an entomologist working on insecticides for mosquitos, roaches, and flies from petroleum derivatives. Lofberg examined samples of tetraethyl for their lead levels. She was trained as a chemist and bacteriologists at Cornell prior to arriving at Standard Oil. Also, mapmaker Lowe Blackman prepared maps from the work of geologists and engineering. She was trained in geology in college and worked for the Army Core of Engineers (Glamour 1945). She was part of the Little Rock Engineers Club, but the membership did not include many women (Glamour 1945).

The oil industry after World War II was opening itself and trying to attract women workers, who could access education, to take technical and other types of jobs in the oil industry. But it was unclear how high a woman could rise in the industry. The industry was starting to open to the idea and started featuring arguments for management training for women in trade publications.

7.5 Making Space and Equity for Women in the Gas Industry

In 1952, Vera Green, who worked as the Assistant Treasurer and Secretary of Botwinik Brothers published an article in *The American Gas Association Monthly* that argued for a larger place for women in business leadership in general. The training of women for management is called "up-training" (Green 1952). Green lays out the reasons that women have entered the work force, and almost all of the reasons involve the desire of women to make money, and not exclusively to support a married household. The editors of *American Gas Association Monthly* editorialized the article by including the leading heading that, "Encouraged into the business world by home labor saving devices and economic need, women have proved able workers[.]" (Green 1952, pg. 11).

Green argued that all aspects of the business world is going to be dramatically affected by the rise in female employment. Working women are also going to have more "purchasing power." (Green 1952, pg. 11). The working woman is also different from the housewife. Businesses will have to adapt to high number of women consumers and producers: "More and more, those of us in business will be working with women, as associates, as consumers, as producers" (Green 1952, pg. 11). She then highlighted the importance of understanding working women in order to improve business, and then Green cited research from census data and found that there were several categories of working women. Some of the working women was made up of non-married women under the age of 25, whom Green states will leave work through marriage. Other women are careerists, who are married, and some are mothers as well. There is also the single woman who is not married, who continues to work, and there are other women without children, who also are married, and who are also working. Finally, there are divorced or widowed women that are in the work force. These women work for "economic necessity" according to Green, because life is expensive, even for married women (Green 1952, pg. 11).

Green cited history, that women "worked hard" to support families through unpaid labor, but then women's work was moved to factories and outside of the home (Green 1952, pg. 11). Women moving into the work force caused complaints from other workers who feared increased competition. However, she dismisses the idea that women should not be involved in the work force, "It's about time they realize that women have not gone out and taken away jobs that were formerly men's but that they merely followed into the public place of work, the job which had been performed at home" (Green 1952, pg. 11). Green argued that women needed work because of their own needs, and also for their families.

Green argued that women should not be judged worthy of a job by their needs, but whether they are competent workers who can perform the job. She used an analogy that the sex of the worker did not matter: "...you know they've never yet invented a machine that could determine the gender of its operator" (Green 1952, pg. 11–12). She also engaged with arguments against the hiring of women for management, the main argument being that women were "'unstable." Green unpacked the argument of instability: "...the unmarried young women under 25, who hopes to get married and to quit work, and the married women with children under 15 or 16, who hopes to stop working as soon as her income can be supplied by another source, either through her husband's earnings, or a drop in the cost of living" (Green 1952, pg. 11). Green pointed out that if that group were excluded, there are many other categories of woman who should be trained for management: these groups represent most of the "womanpower" in the workforce.

The candidate for management should be judged by a hiring manager or supervisor and the applicant should be judged fully, "It is essentially a matter of selection. It is a matter of obtaining full information regarding background, regarding the applicant's aims and purposes and desires in life." And the personal manager should determine which group the applicant fits into, a category that was deemed "unstable" and she did not recommend hiring a young woman looking for marriage, or a married woman with kids for a "key job" (Green 1952, pg. 12). Young women, who are looking for marriage, according to Green should be hired in the following way, "Such a girl should be hired into a job where she can be readily replaced, so that she does decide to leave, there won't be moaning and groaning and gnashing of teeth about the instability of all women employees!" (Green 1952, pg. 12). However, career minded younger women, who were thinking about marriage, but prioritized their career, should be eligible for management jobs, Green added to her analysis. Green pointed out the changes to the hiring of women, "We can probably all remember the time when companies simply did not like to hire married women. But today 93 percent of our employers have the welcome mat out for married women and good number prefer them" (Green 1952, pg. 12). At the time of the publication of the article, more women had the status of married, than those who were single. Green also argued that statistical information showed that only two percent of working households had their marriages end.

The married woman, according to Green in her 1952 article, were simply happier: “It is generally conceded that working wives are happier, too, whether they are working at jobs, or some other outside activity, as long as it is a healthy interest outside the home” (Green 1952, pg. 12). She also moralized the working woman, and argued that the husband should support their wives, “Husbands, who tie their wives down and discourage them from seeking normal outlets of self-expression outside the home, are a guilty of wrongdoing as dotting misguide mothers who tie their children to their apron strings throughout their lives. If a woman can work—and wants to—she should be given the opportunity; the prerogative should be hers” (Green 1952, pg. 12). Green also defended the abilities of women in the work force by citing the findings of the Massachusetts Division of Employment Security, “Women will match a man on dexterity, mental ability, and production. The only problem involved is plan physical strength” (Green 1952, pg. 12). Another authority figure Green cited stated that, “Women are just as a strong, brainy and capable as men. They live longer, too, and could have bigger muscle s if they worked on it!” (Green 1952, pg. 12).

The article also highlighted the purported discoveries of the war era, that women could fill the needs of the workforce. Though there was no attribution, Green cited the quote in an unnamed business journal, “One of the greatest discoveries made in the past war was that women are able to do practically anything that men can do” (Green 1952, pg. 12). Green then makes her biggest push for equity for women in business and leadership,

> If we, in business, will operate on that premise, we will have many of our personnel problems solved. Naturally, every woman isn’t going to be able to do every job expected of her, but, then neither is very man. Let’s concede that individual ability is based on something beside sex. We have capable men and we have less capable men. By the same token, we have capable women, and we have less capable women. But when women are involved, their lack of capability is blamed on the fact that they are women, not ascribed as in individual failure. (Green 1952, pg. 12)

Women, according to Green, should be treated as individuals, much like men in business, and not be essentialized by their sex. She pointed out that women are better educated than ever before and ready for more business opportunities. But she concedes that though all women want to be successful, many will not advance beyond their entry level jobs, or travel to higher administration in companies. Green pointed out that this is the case with many men as well. Education, Green emphasized, is preparing women to advance: “But, as women enter the higher educational institutions in greater numbers, more leadership will be developed among them. And it is this very increase in educational training for women that is enabling them to work their way into management jobs” (Green 1952, pg. 12).

She historicized women working along the same years that this book takes place, from the 1900s until the 1950s. Green pointed out that few women were in college, but now there were more colleges that accepted female students than men, as there were at the time 273 colleges for women only in the United States, and 238 that only accepted male students. And women’s colleges have significantly revised their curricula that prepared women to succeed in the business world. Women have pursued many educational opportunities as whole. They were succeeding at universities, as she pointed out that the overall grade point average for female students at the University of Wisconsin was higher than that of male students. High achieving students at that university were elected to Phi Beta Kapp, and almost twice as many women as men were elected to the academic honor institution.

Green called for more support for women, as women become more and more educated, compared to their male peers, and there needs to be more opportunities for those women to fill. Women needed more circumspection about what their career trajectory might entail, Green argued, and she noted that women missed out on career planning, especially early in their career,

> Today, as our needs grows for women supervisors, for women to fill higher level jobs, we find a noticeable lack of specialized training and technical knowledge from women already on-the-job. This could be attributed to two causes: The whole idea of women specialists is a new and our present working women were not alerted to this need early enough to do anything about it while in school., And second, there has been a hesitancy on the part of companies in furnishing training for women employees. (Green 1952, pg. 12)

Green was likely unaware that women had started out as specialized labor in the oil and gas industry, and it was the non-specialized jobs, along with specializations, that had expanded for women. Also industry and the US government in publications reviewed in previous chapters of this book also pointed to a more knowledgeable female worker in petroleum geology.

The stakes of not addressing these problems for female workers were high, and potentially harmful to industry, as there would be a "severe shortage of office supervisors" as companies expanded and male workers left industry to serve in the Korean Conflict that had started in 1950, and ultimately 1.5 million men were drafted into the United States Armed forces to fight in the conflict that remains ongoing to this day (Green 1952, pg. 12, 37).

Office supervisors were ignorant of women working in management, as Green pointed out that there were women serving in management in almost every type of business. The largest population of women working in management was in banking and insurance. Green also examines some of the stereotype that women experienced in attempting to gain management position. The idea that women did not like to work for female managers was addressed by Greene, who dismisses this by saying that there are bad female supervisors, just like there are bad supervisors in general, but these assumptions about women not working for women is "old fashion" and does not fit the current realty of office culture (Green 1952, pg. 37).

Green advocated for a moderate attitude to being a woman and feminine in the work place. She acknowledges that some women try and dress like men, even "cuss" and "talk" like their male counterparts. The other extreme Greene mentions is that they try to be extremely feminine: "Others go to the opposite extreme, wear plunging necklines to the office and too much makeup and murmur ingratiatingly to their assistances, 'Darling, would you mind terribly doing this for me?' Nobody likes them either" (Green 1952, pg. 37). New executive women are mostly liked by their employees and they have one "common denominator" and that is that "They love the work they are doing. Because they love it, they think in terms of the work itself.

And they conduct their offices on the basis of know-how, not on the basis of sex. They have advanced in their position through hard work and the willingness to give more than full measure on the job" (Green 1952, pg. 37).

Green continued to discuss the old and the new working woman. She said that men, and perhaps women, calibrated their expectations of the new "modern business woman" by mixing the old and the new, "She is a combination of the old and the new. She is the old-fashioned gentlewoman who has retained her charm, her femininity and her fastidiousness; but who has added a degree of smartness, a flow from the outdoors, a freedom of activity, an analytical mind, an indisputable sense of humor, and the 'ever-new' look of fashion" (Green 1952, pg. 37). She sums up the quote by emphasizing that the women who are successful in business are fashionably dressed, maintain courage, clever, and are "poised" (Green 1952, pg. 37). After listing the multitude of characteristics, Green admits that it is a difficult list, but women must have them all.

In the most "progressive" industries top management has seen the value and need for women in the highest positions. Green gave examples where excellent female managers improved the morale of the company, compared to the previous male occupant of the position. Green also cautioned that women in offices not working well with female managers risk their careers. She even addresses male resistance to female leadership, and found that men are likely becoming more accepting,

> Men in business today, actually without realize it, are thinking much more broadly regarding women in business. Maybe there are so many women in executive jobs now they're getting used to it, or maybe they feel instinctively the trend that is taking place, and know that refusal to accept women on an equal status in rapidly being classified as another form of isolationism. (Green 1952, pg. 37)

Green ended her argument that there needs to be an end to the "prejudice" against women in management. Minds being open to female leadership will make the world a more improved place, and she called for the reader to contribute to the construction of that type of world (Green 1952, pg. 37). Again, a trade magazine like *American Gas Association Monthly* running this type of article, with this type of argument, was seen by industrial culture as an important one for managers to entertain.

Women were entering the oil industry in geology, and related jobs, in managerial, or executive administration. One example can be found in the self-confident and exciting profile of Rosalie Armistead Higgins. Higgins was profiled in the St. Joseph News-Press in 1953 and believed herself the equal of any man (Robb 1953). The New York based profiled opened that, "As a staunch believer in equal rights for men, I hate to see the boys being crowded in any one of the numerous business and industrial fields they long ago talked out 'for men only'" (Robb 1953). Higgins pointed out that women were moving into insurance and the oil business. And she had great confidence in recommending them for women seeking employment, "Both fields are wide open for women with git-up-and-go, with the only limit the extent of the girl's abilities" (Robb 1953). Higgins likely shared many of the beliefs that Green expressed in her magazine piece just a year prior to this newspaper article.

Higgins' photograph might appear in the piece, though the image is unnamed (Fig. 7.7). Her image is strikingly confident and does not include any men or symbols of her business. This is a drastic break from the pin-up style images in World War II. The description of the "wrap-around coat" is also interesting because it was designed to take up a larger amount of space. Women, in the United States and Europe, are often encouraged to take up as little space as possible, and diet culture

Fig. 7.7 This might be an image of Rosalie Armistead Higgins (Robb 1953)

pressures women to take up even less physical space, as many scholars have pointed out in several studies (Matelski 2017). The coat is described as big, that can be made bigger or smaller, depending on the needs of the woman.

Inez Robb, was a female reporter, who wrote this piece on her neighbor, which was also a break from the mostly male centered reporting during World War II. Robb included confident descriptions of Higgins, "Women are natural-born insurance salesman because so often the problem of insurance hits so close home to us" (Robb 1953). Higgins published a book about the importance of insurance that had gone through two editions. Formerly working as a newspaper reporter prior to selling insurance, she noted that her book was read worldwide, in such places as the Philippines and Turkey, as well as at Harvard University. She, and Robb, argued that the success of the book was because of the "feminine" approach that was as they defined mostly "practical" (Robb 1953).

Higgins remained active in the business world because of her experiences in administrating several business clubs for women in New York, and argued that women could make a very good living selling insurance.

Robb also profiled Ernestine Adams, who she knew from her time in college at the University of Missouri (Robb 1953). She was named as "Oil Woman of the Year" and she optimistically claimed that "The sky is the limit for women in the oil industry[.]" She was editor of *The Petroleum Engineer*, a major trade magazine that was discussed in this book as it often featured working women in the oil industry. Her work as editor was contrasted with the past, mostly male, editorial leadership, and her ability to enter into "…a masculine stronghold" (Robb 1953). She proudly promoted the openness of the oil industry for women. The article quoted the help that the oil industry offered to women, "'The men in this industry will give a great, big hand to any woman with ability who'll work. She can aim as high as she pleases, and get there, too" (Robb 1953). She also promoted entry, as well as higher paying jobs, readily open to women in the oil industry,

> The sky is the limit for women in the oil industry, …There are thousands of executive secretarial and office management positions for women in the oil industry,… But for women with specialized training, or who are willing to take extra schooling, there are wonderful jobs as geologists, engineers, draftsmen, geophysicists and petroleum chemists. (Robb 1953)

Adams then recounted a personal example of woman achieving success in the oil industry in South America. She promoted the story of a female research geologist prospecting for oil, "I know a woman who runs a successful scouting service and another who is a research paleontologist. Look at Elsa Pierce who is assistant to the manager for the Venezuelan manufacturing department of Phillips Petroleum in Caracas. Venezuela" (Robb 1953). Higgins then tried to joke with the readers that they did not to pursue marriage, for instance to Howard Hughes, the wealthy Lubbock born drill bit heir in the petroleum business, but "…it would give a girl a head start, or make a fortune in insurance by conking out a husband for a billion dollars" (Robb 1953). She emphasized that making a living in oil was a very accessible career path: "It can be done just as readily with perseverance and perspiration" (Robb 1953).

Adams continued to learn about the oil industry, including taking mathematics classes related to the oil industry at Southern Methodist University (Golding and Russell 2022). She also continued to be one of the foremost champions of the oil industry, publishing the essay, "What's Wrong with Being an Oil Company." The Essay was read into the 1951 Congressional record, which argued that oil companies the United States, though often criticized, produced good wages for their workers, opportunities, and sufficient supply for consumers in the United States, as opposed to Russia or other "socialist" nations. It is surprising when Adams was praising the opportunities and education of workers, she did not directly refer to women. Adams was recognized as "Oil Woman of the Year" for this essay in 1959 (Golding and Russell 2022; United States Congress 1951; The Leader-Post 1959).

Though women were promoting the openness of oil companies to female workers, what was the day-to-day culture like for women at these businesses and how did they engage these female employees? Oil companies wanted to engage and build female culture with their employees. Discussion groups internally were another aspect of office culture and female participation in oil and petroleum geology. Discussion groups brought women the latest information from their own industries and served as another focal point of education, and community that women actively built, though limited by corporate male interests. Discussion groups served as another example of the new woman executive, but influenced by older cultural ideas, such as proper office attire.

7.6 Female Geologists in the Office: The Mobile Women's Discussion Group

Oil companies wanted female participation in the oil industry in the postwar era, and both companies and women were optimistic about the potential for the openings in this field (Coll 2013). Mobil Chemical Company established "Women's Discussion Group," in the 1960s (Women's Focus Group 1951–1968/1962–1963). It seems that Mobile wanted to engage with its female employees, and went beyond simply recruiting women to fill jobs. Standard Oil and Mobile Oil later merged to form Exxon-Mobile, so the interests of one company likely speaks for the other, as Standard Oil (later forming itself into Exxon) clearly showed an interest in female employees (Coll 2013).

By examining the Women's Discussion Group, we can get a perspective into how these companies had a culture that involved female employees, and also how active and involved women in both the professional and clerical positions in petroleum geology were in forming their own identities within the companies. Most of these sources comes from notes, memos, and intra-office communication, so these sources are a best representative, and certainly not comprehensive of the experiences of women within the Mobile Oil Company. I think with a certain historical perspective whether someone in the future read my career through memos, emails, company contests, and promotions, but these are the sources that sometimes remain in the historical record. These records were presented to the University of Texas, so potentially records could have been destroyed, accidently omitted, or gone unpreserved by the donating party. In historical language, we call this preservation bias, as we only know things about the past by the records that are kept, preserved, or are simply not destroyed (Women's Focus Group 1951–1968/1962–1963). However, these internal documents speak about the culture of the company and how the Women's Discussion Group would operate in the company and involve female employees.

The Women's Discussion Group, which was established in 1951 in order to facilitate familiarity of employees of Mobil Petroleum Company and "…for the purpose of learning more about our Company and industry," as the company was merged in 1962 with the "Mobil Family" (Women's Focus Group 1951–1968/1962–1963). The memo was sent out to "LIST 'G'—ALL WOMEN EMPLOYEES- Socony Mobil, Mobil Chemical, Mobil International, Mobil Oil, Mobil Petroleum… "(Women's Focus Group 1951–1968/1962–1963). The group would meet once a month, with "light refreshments[,]" and "All women employees are invited to join." The group was led by Anne C. Rouse, who served as Chairman; Helen V. Doherty, as the Deputy Chairman; Beatrice Cunningham was the recording Secretary; and Florence C. Murray served as the Corresponding Secretary.

In 1966, the Women's Discussion Group wanted to have a contest to name itself. The winner was "Mobilogue," combining the words Mobile and Dialogue, and the creator was Beverly Diamond. But it seems that the company's legal department had a problem with the use of Mobile from a group that was not technically associated with the company, while the memo literally referred to "ALL WOMEN EMPLOYEES" and included the groups "Corporate, Chemical International, and North America." The problem with "Mobilogue," as described in a memo "TO ALL WOMEN EMPOYEES OF MOBIL OIL CORPORATION[,]"

> Then, and only then, we discovered that legally, the name "Mobil" cannot be used as part of an official name for an employee group, which is technically as "outside group." To do so might result in weakening the company's trademark position. This was unthinkable to a group with our company loyalty and interest, so we had to forget about not only the winning name, but many of the other really descriptive suggestions. (Women's Focus Group 1951–1968/1962–1963)

Many people associated with a large institution or corporation has had a similar, morale reducing experience. The officers of the group, both in the past and current, gathered together to come up with a new name, and decide to award winners, including Beverly Diamond, who received a Mobile Travel Guide. Some options were recognized as winning, and also receiving Mobil Travel Guides, included "Distaff Forum," "Mobile Distaff Club," and "Mobil Femme Forum" (Women's Focus Group 1951–1968/1962–1963). Distafforum was chosen because of its many meaning that were promoted in the memo.

The memo went on to explain that the work "Distaff" has a meaning of women or collective women. The definition of forum is also explained, that would have likely been clear to the memo reader, but nonetheless defined it anyhow. The selection of forum was thought to capture what the group did, hold public meetings and attend lectures. The signers of the memo, all women, promoted the news name, "The more we think about DISTAFFORUM, the more we like it. We hope you do too" (Women's Focus Group 1951–1968/1962–1963).[1]

The newly named DISTAFFORUM, noted as "formerly Women's Discussion Group," met from 1951 to 1968. Listed on the notebook that discussed their activities and was donated to the University of Texas as their historical record contained a note, "Betty—…I weeded through and tossed all/the 'important' papers./Believe attached is worth keeping./Elena" (Women's Focus Group 1951–1968/1962–1963). Unfortunately, there is no telling what type of historical records, events, activities, are missing from the historical record.

Beatrice Cunningham noted that on the meeting of September 140, 1962, the Women's Discussion group had 119 women joined the group, and 105 women attended the meeting. The original speaker, a Mr. James Christensen, did not attend the meeting and Eleanor Dickson was able to substitute, with favorable reviews on September 12, 1962. Eleanor Dickson was sent a type of thank you memo with positive comments shared with her: "'What a nice personality she has,' and many more of the same" (Women's Focus Group 1951–1968/1962–1963).

Eleanor Dickson was the "Socony Mobile's Secretary Department" and she made a presentation about Japan. Her background was given in the introduction to her September 7,1962 talk by the officers of the group. Dickson started working at the Sandard-Vaccuum Oil Company as a stenographer in accounting in 1935, and then advanced to become a legal secretary, and continued her education at New York University. She was ultimately promoted to the secretary of the corporation and worked with its board of governors, as the only women in that capacity. The introduction also included information about "Miss Dickson's" hobbies and her extensive personal and professional travels in Japan. Other activities of the focus group included a "tour of a Company installation[.]"(Women's Focus Group 1951–1968/1962–1963). Meetings of the Women's Discussion Group continued in a regular and consistent manner with invited speakers giving presentations of general interest and industrial topics, and all of the women at Mobil were invited. Many topics were shared, including information about new explosives and new ways of producing energy.

On October 1, 1962, the Women's Discussion Group organized more of a social meeting to build community. The memo attached to the announcement of Dr. Robert W. Schiessler reminded members of the group that "Part of the purpose of the Women's Discussion group is to meet your-co-workers[.] Your Committee has given a lot of thought about ways we could arrange for more introductions, and here are some suggestions that may help all of us" (Women's Focus Group 1951–1968/1962–1963). The memo reminded members to share meals in the cafeteria, and join others eating that you might not know yet at the company. Other topics included the presentation "Operations on the Life of Soviet Women" from a speaker that worked at the Arabian Oil Company, and a presentation about oil

[1] See page 31 for a mention of Connie Moon, (Effective communication for engineers 1975).

drilling and exploration on the coast of Louisiana (Women's Focus Group 1951–1968/1962–1963).

The Women's Discussion Group acted like a type of Desk and Derrick club for female oil employees, having both technical and social components. The Women's Discussion Group was likely seen by employees and even management as a necessary component to build good culture at Mobil. However, ideas about women, especially their proper place in the company, were additionally spread through periodicals and human resources memos in the 1970s responding to and discussing proper attire at the office. However, the Women's Discussion Group was an aspect of corporate culture mostly controlled by female actors. Unfortunately, many of the records were not preserved and we only have an understanding of the topics, organizing principles of the group, and some brief information about meeting attendance.

7.7 "Good Judgement": Women, Office Culture, Misogynism, and Fashion

In 1971, Mobile circulated a memo discussed appropriate attire for women in corporate offices (Exxon Mobil 1917–1985). The memo attempted to discuss the changing nature of fashion and the Mobil company's approach to changes it implemented prior to the memo in the summer of 1970. The memo indicated that some supervisors had problems with the apparel rule changes brought on by allowing the "pant suit" (Exxon Mobil 1917–1985). The historicization, and subsequent retreat was contained in the memo, "This past summer we approved pantsuits as appropriate attire for women for office wear at Mobil. At that time through individual supervisors, it was indicated that pantsuits are considered pants and a matching or coordinated tunic or jacket, and we requested that your employees refrain from wearing blacks and sweaters or blacks and blouses" (Exxon Mobil 1917–1985). The memo attempted to convey liberty to women in the workplace, they reminded employee of the company's ideas of proper decorum, "We want Mobil women to have every opportunity to enjoy the new fashions, however at the same time, we would like them to use good judgement in maintaing a respectable and business-like appearance in their office attire" (Exxon Mobil 1917–1985).

The memo conceded that this had been achieved but the memo needed to serve as a reminder as some employees were breaking with the opinions of the company. The memo defended the reminder as "For the most part this has been done, but of late we have noticed some modes of dress we do not consider appropriate for a business office." And, the memo added that the restricted clothing items applied to "…the guidance of our female employees, we are listing below dress that is considered inappropriate [underlined in the document] for office wear at Mobil" (Exxon Mobil 1917–1985). The inappropriate items are listed in the Table 7.1 below:

Table 7.1 List of inappropriate clothing at Mobil (Exxon Mobil 1917–1985)

Slacks with sweaters or blouses	Jeans or Denims
Culottes	Knickers
Jumpsuits	See-through fabrics
Shorts	Extra wide or flared pants bottoms
Gauchos	Tank tops, pants, etc., that fit too snugly

No guidance was given to male employees, but the women were thanked in the memo. All the memo went out to LIST G, which was all women at Mobil. The changes in such as memo contrast with newspaper articles and stories, even in trade publications like the 1944 article "From Dimity to Denim" that appeared in *The Petroleum Engineer*, or stories that embraced the field fashion of June King (The Petroleum Engineer 1944). There appeared to be a backlash against female employees in the 1970s.

In the 1960s–1980s, women's office and school apparel were a point of cultural discussion. The second women's movement was taking off in the 1960s, and feminists and women interested in changing fashion trends challenged the conventional dress of the 1950s. However, as the reader has seen from reading this book women geologist wearing clothes of rough-neck oil well wildcats were equally praised for their hyper feminine dress and movie start looks. The 1960s with the advent of miniskirts and hemmed dress lines that rose caused some discussion in corporations and among office managers. However, the interesting cultural point to be made in the list on inappropriate items mostly focus on women wearing pants and the pant suit, not the length of women's clothing.

The Canadian newspaper *The Calgary Herald* discussed the questions related to the modern working woman shown in Fig. 7.8, "Mod Clothes Out For Job Hunters" (Stratas 1967). Diane Stratas summarized the dilemmas women faced by interviewing, in a false persona, for open jobs. The expectations of hyper femininity was difficult, as women had pressure to dress fashionably, but try and guess what job interviewers had for their expectations and the corporation. "Susan," who was based on Diane Stratas but her take persona that dressed in a miniskirt, dangly earrings, and net stockings, the standard of the 1960s, found difficulty with most interviewers, such as an interviewer who said to her that, "'The miniskirt will never be acceptable in business here. And let's face it Susan, men are distracted by short skirts,' he said from behind his hanky" (Stratas 1967). She promptly left the interview.

She found that the problem with skirt length is not from employees or interviewers, it is from executives in charge of corporate image, pictured in Fig. 7.9: "Girls like the mini, men like the mini, but management is shutting them out to preserve the corporate image and not offend 'the public'" (Stratas 1967). She also talked about women had limited options into applying for jobs where the office culture does not welcome them, "A girl who spends 16 years at school doesn't study shorthand and typing and have at least five years' office experience is over-qualified and under-experienced for job market. And she is even in a tighter fix if she wears a miniskirt" (Stratas 1967). Managers explained that though they had no problem

Fig. 7.8 Newspaper advertisement for women's clothing (The Tuscaloosa News 1972)

with the skirts, they worried that older clients might wonder about the credibility of the institution, or how "frivolous" younger persons were hired at such an institution.

She also mentioned that even though she had a university degree with honors, most employees were wanting a woman with an "applied science [degree.]" She often found she got the question, interviewing with a pure science degree, of whether she was going to stay at the business: "Are you sure you don't want to teach school or go back to university, Susan?" (Stratas 1967). She was asked this question at three job interviews. She also commented on the harassment she got from passing motorists, men on the street, and older women, and business men gawking.

There were even articles that expressed misogynistic office manager fears in the 1970s, such as the article with the headline "Legs Vanish As Pant-Suits Invade Canadian Offices" (The Calgary Herald 1970). In the joke section of *Oil Week*, entitled "Crude and Refined," a joke ran that summarizes attitudes of some people toward the non-femininity of women who wear pants. The joke read that, "Women who insist on wearing the pants frequently discover that it is the other women who

Fig. 7.9 Article explaining the controversies about miniskirts (Stratas 1967)

are wearing the mink." The joke implied that if a woman decided to wear pants, she would miss out on the material rewards of a romantic relationship (Oilweek 1974, pg. 28).

The jokes in the section continued to convey themes of rape culture and misogynism in both American and Canadian business and homophobia (Baker 2004; Elias 2022; Harkess 1985; Harrington 2020; Jackson 2023; MacKinnon 1979; Osgerby 2001; Johnson 2004). The jokes appeared next to announcements about professional meetings and was a major trade publication that also ran job ads. The jokes focused on alcohol and consent, "Martinis, my girl are deceiving: Take two at the most./Take three and you're under the stable./ Take four and you're under the host." While others implied that women's bodies were open for male access: "A girl with an hourglass figure can often make grown men feel like playing in the sand" (Oilweek 1974, pg. 28). Attacks on homosexual women also appeared: "Definition of lesbian: A mannish depressive with delusion of gender." Other jokes included crude sexual references: "What's worse than a piano out of tune?/An organ that quits in the middle of a piece" (Oilweek 1974, pg. 28). These jokes appeared around 1974, which was around the same time that Mobil was enforcing clothing codes in the office, and is likely representative of the male culture of administration.

A journalist writing about attitudes regarding corporate dress, Ron Scherer, wrote in an article that originally appeared in *The Christian Science Monitor* that by the 1980s, corporations, even very conservative corporations like International Business Machines (IBM), were not concerned with short skirts making a return to the office. Scherer historicized the changes of the 1960s,

> Unlike the late 1960s, when some corporations tried to establish dress codes for employees, the fashion trend toward shorter skirts—some only a few inches above the knee, some that only come down to midthigh—apparently hasn't resulted in new corporate dress standards. (Scherer 1982)

Scherer interviews "spokesmen" from companies like IBM and Exxon. His comments were informed to a writer at *Working Woman*, a magazine that covered women's experiences, challenges, and trials in the workplace, removed the onerous codes of prescribed inches of length and violations to more nebulous comments with vague, unclear meanings. Marilyn Machlowtiz, of *Working Women* informed Scherer that, "the prevalence of plants made policy statements on dress passee[,]" and put in their place statements like "we expect all employees to maintain an acceptable dress standard" (Scherer 1982). A representative from Exxon reflected the same idea in defining a company dress code for women, as the article reported that of Exxon that "...the company has no formal dress code, but expects its employees to 'dress with taste'" (Scherer 1982). The article goes on to further report why women were wearing skirts of different lengths beyond contemporary fashion. Some workers stated that shorter skirts were cooler and more comfortable in 80-degree climate in summer seasons in New York, while some simply associated it with "freedom" (Scherer 1982). One of the most judgmental comments that came from Marjorie Gaber, who worked in sales at Geraldine Peterson Sportswear, implied that if women wanted to rise in the executive structure, they should dress more conservatively saying that: "the miniskirt is not back to stay. And, she notes, the executive-caliber woman will probably never wear one to the office" (Scherer 1982).

The debate about women fashion continued in the 1980s and internally, and stories from New York and Associated Press writers appearing in Canadian newspapers. An article by Elleen Putnam ran with the headline "Executive women now sport feminine attire" (Putnam1983). The article summarized that "Women in the executive suite are finally learning that it's OK to look like a woman at the office." And that colors and contemporary fashions were acceptable. "Dresses, frilly blouses and softly structured suits are becoming fashionable for women executives who used to feel comfortable only in female versions of the traditional male principles, grew flannel or oxford cloth" (Putnam 1983). The writer of a book that guided female executives through fashion choices, Mary Fiedorek and her book *Executive*

Style, punctuated the situation by pointing out that there were simply more women working and that provided more freedom to choose what to wear (Friedorek and Jewell 1983).[2]

The pant suits that were so cutting edge and boundary pushing to executives at Mobil in the 1970s were seen by the new tastemakers of the 1980s as conservative and boring: "And while suits constitute to be the primary choice of upwardly mobile women, they are no longer designed to disguise the female form" (Putnam 1983). Fiedorek summed up the changes that had occurred since the second women's movement: "Fiedorek said she believed that while the feminist movement exposed sexism at the office, it also left women feeling that to be accepted professionally and avoid harassment they had to look like men or be dowdy" (Putnam 1983). Most representative of the significant changes suggested by Fiedorek, whose opinions were then suggested to the reader as a matter of proper taste, was that the reader should basically leave the A-line skirts of the 1950s for the dirndl, or wide statement making dresses and other clothing that emphasizes "softness" (Putnam 1983). Most of the women, who were working as female executives making $40,000 a year and were around 32 years of age, were firmly against pants because they felt that they made them too "masculine."

In historicizing these events in fashion and women working as petroleum geologist, the changes that occurred in the 1970s and early 1980s are different than those profiles in oil magazines during World War II. The profiles written during World War II and shortly after the war were composed by men, but now in the newspapers of the late 70 s and early 80 s, women are having more of a voice into their thoughts and opinions about clothes, and they are less and less enmeshed in male opinion, as the pin-up write ups emphasized the beautiful woman in a man's clothes (Reichert 2003; Chrisman-Campbell 2022; Hillman 2015).

Women interested in geology were not absence from changes and debates surrounding women's fashions. Louise Mitchell was profiled in a write up of the women married to the Apollo 14 astronauts. Mitchell was, at the time of the Apollo 14 mission in 1971, enrolled as a graduate student at the University of Houston, pursuing advanced education in special education (Bangor Daily News 1971). Mitchell, defined as the wife of Edgar Mitchell, was described as "shy and soft-spoken." Her appearance was also described as wearing a "plaid miniskirt and white boots" (Bangor Daily News 1971). She had undergraduate education in fine arts at Carnegie Tech, but studied geology intensively in junior college, and has just recently finished her bachelor's degree. During her time studying at San Jacinto Junior College, she had taken several science classes including geology. She remarked in the interview that "'I loved geology.' She said [,] 'I like rocks better than my husband does'" (Bangor Daily News 1971).

Much changed between the lifetime of Augusta Hasslock Kemp and the 1940s. War and postwar fashions changed dramatically, and women increasingly

[2]The book Executive Style: Looking It, Living It, has a cover that credits Mary B. Fiedorek "created" the style, but "Diana Lewis Jewell wrote the book (Friedorek and Jewell 1983).

advocated for more freedom to wear what worked well for them (Hillman 2015). Though women were pushing cultural and gender boundaries with new fashion, culture in some oil companies reflected the pressure on women to fulfill traditional domestic roles.

7.8 Mobil Pushes Standards of the Housewife in a Story Entitled "Are You a Handicap to Your Husband" and Readers Respond with Frustration

In 1957, an article appeared in *The Magnolia News*, which was a publication produced by Mobil. Specifically, the publication was "…published for the employees of Magnolia Petroleum Company and Magnolia pipe Line Company by the Public Relationships Department."[3] The article "[A]re you a handicap to your husband?" ran in *Magnolia News* in the July–August 1957 edition of *Magnolia News*. The article led to a lot of letters to the editors of the magazine, causing the mostly male editorial staff to write a rebuttal entitled "Who's a handicap, you bum?" The editorial staff was made up of H. T. Fort Jr., Harold W. Hoffman, Joe T. Arnett., and Charles F. Morrison (The Magnolia News 1957b). The first lines of the rebuttal summed up the editorial response, "That article we ran on 'Are You a Handicap to Your Husband?' provoked some wrathful letters from employee's wives—in fact, more letters than we've ever received on a single article" (The Magnolia News 1957a). The editors noted that no man wrote any letters about the article.

"Are You A Handicap to Your Husband" involved employees at Mobil, such as Gus Kester and Clar and Kay Coleman. The article opens with Gus Kester, fresh from a hectic commute, and having to force a positive attitude when he arrived home. The article then switches to summarizing what the article will criticize as a bad home dynamic of a wife, presumably Kay Coleman criticizing her husband. The article jumps between Gus Kester and a man, pictured with Kay Coleman named Clare. The article is unclear which man it is referring to, so perhaps Clare is Gus Kester's middle or nickname; this is not specified.

Clare has a lot of pressure on her as homemaker. And the article puts all the responsibility of the household on the woman. The statements that Kay Coleman makes in the article are would be best characterized in the stereotype of the "nagging wife." The stereotype of the nagging wife often involves a woman criticizing her husband, despite his best efforts, to manage household chores (The Magnolia News 1957a). Historians and social analysis today would characterize the behavior of such a man today as "weaponized incompetence."[4] She asked the husband Clare as soon as he arrived, "Where have you been? You're ten minutes late and you have

[3] The original was italicized.

[4] The term has a pop-cultural association with it, and I could not find much history on the term. Please see: https://www.psychologytoday.com/us/basics/weaponized-incompetence

to go to the store before we can eat" (The Magnolia News 1957a). Then he is given another set of tasks to complete, and responds to his wife that "O. K. O. K. Just a minute, hear. Let me rest my weary bones a second first" (The Magnolia News 1957a). Then the criticisms and frustrations continue to flow out: "Rest nothing. I haven't had a rest all day so why do you need one? No! No! Don't sit on the couch. Can't you see I just cleaned the cover?" (The Magnolia News 1957a). Clare continued to be painted as a sympathetic figure, and Kay Coleman continued to criticize her husband. Gus [Clare] eventually took the children to the grocery store and complained the whole time while completing the errand.

The article stresses such a home dynamic as bad to productivity and a threat to safety on the job. Gus [Clare] works for Mobil and has problems at his job because of his relationship: "At work he never quite seems to have his mind on the job. He goes through the day just plodding along with no enthusiasm at all. Gus never seems in too much of a hurry to get home, either. Can't really blame him, can you?" (The Magnolia News 1957a). The article then goes on to imply that the housewife is the invisible worker who contributes to a positive work environment. However, note that the housewife has all of the responsibilities of a work, in addition to the family, and is not paid. The article noted that companies like Mobil are paying attention to these hidden hindrances to productivity, and the nature of an employee's homelife. Mobil again, puts all the pressure on the housewife,

> Industry today recognizes the important role of the housewife plays in a man's work as well as his home. If a housewife behaves like Kester's wife, then her husband won't be very happy at home. Such a wife becomes a handicap to her husband rather than a help. Because if a man isn't happy at home, he won't be happy at work. (The Magnolia News 1957a)

The article was quick to note that if the home continued to be bad, productivity dropped for works: "If an employee becomes unhappy, his efficiency usually drops" (The Magnolia News 1957a). Then the article continued to describe the "vicious cycle" for male works. The importance and pressure on woman to produce a proper home life is highlighted in the article, "This prevents him from receiving the recognition and job satisfaction he deserves, which in turn can make his home life even more unhappy. It's a vicious circle. But it works in reverse, too. A happy man at home will be happy at work and this in turn adds pleasure to his home life" (The Magnolia News 1957a). The home must be a site of love and understanding, and the product of women's work: "If done through love, understanding and sympathy, rather than ambition and selfishness, a woman can help her husband be a lot safer and much happier at home and at work. She can see to it that he is happy and in turn, her own life will be more rewarding" (The Magnolia News 1957a). A safety director also concurred with the statements about the relationship between home life and on the job safety.

The article then recounted some bad experiences for workers that resulted from their homelives. One employee, Charlie Louis, after getting into a fight with his wife, had a traffic accident on the way to work. In oil derricks, Bob Graham forgot to secure the elevator that lifted drill pipes, and this mistake resulted in the injury of another oil worker. Graham, it was noted, had been arguing with his wife about issues of household cleanliness: "The night before this happened, Bob and his wife had a row about sitting on the furniture in his work clothes" (The Magnolia News

1957a). And the blame for the workplace accidents did not fall on the responsibilities of the workers. Instead, the article argued that, "These accidents would have been prevented if Charlie and Bob had had their minds on the job and not at home" (The Magnolia News 1957a).

The image of the wife as "positive thinking" was stressed as crucial to making a safe and comfortable household, which prevented work place accidents. And wives were blamed for the stresses of male employees: "A troubled mind causes a man to be careless and make errors in his work" (The Magnolia News 1957a). Even one of the company doctors stressed the importance of women and household management with that of employee health: "The woman of the house should insist that her husband take care of his health, both mentally and physically" (The Magnolia News 1957a). Not providing husbands with complete meals and simply coffee endangers workers and the industry, and women should understand the proper principles of diet. The principles emphasize restrictive eating and fatphobia (The Magnolia News 1957a; Wdowik 2017; Matelski 2017; Doan 1961). The housewife must also make news friendships for the family and make most of the decisions in the household, and pre-emptively prevent problems for her husband.

The article cited the latest health information to justify its arguments. Citing the work of psychologists, 700 married couples were surveyed in an unnamed study, where husbands complained that their primary criticism of their wife is "...nagging, whining, and sneering..." which the study claimed "...causes anxiety and mental uneasiness in the husband" (The Magnolia News 1957a). The article then offered methods to remove nagging from the marriage, such as the woman being more gentle, minimizing or laughing off issues, and more calmness, and the housewife should never raise her voice. Raised voices caused "ulcers" in husbands.

The article concluded that the housewife had a calling to make the household calm and "a haven," where the housewife took on many roles. However, none of these rolls were paid. And the article then moralized all the responsibility of the housewife married to an oil employee,

> No one will say that a wife's job is an easy one. You must be a guardian angel watching over the health of your family. You must be a nurse, a morale builder, a sympathizer, an organizer, a dish washer, and a baby sitter. You must be the hone who makes a house a home. You must be the one who sees to it your husband and your family have a happy life and a safe life. But what do you get in return? The things you want most in life...lover and recognition!". (The Magnolia News 1957a)

The article did not list any responsibilities for men.

The article ended with a checklist for housewives to determine if they were "handicaps" (The Magnolia News 1957). The editorial responses to the implied high volume of letters that were written into the magazine by female employees, which could have included everyone from geologists to office workers, were dismissed. The editors claimed that the article was so interesting that an unnamed nationally syndicated publication wanted to have a copy for presumed republication. I was not able to find any republication in magazine and newspaper searches.

The editors also claimed that some housewives liked the article, especially "...those that made high scores on the quiz" (The Magnolia News 1957b). But that was the end of addressing criticisms in the editorial. In December of 1957 a female editor, Mary Elizabeth Lewis, was added to the legal department. She had worked her way up in the company from an initial position as a stenographer, and as a practicing attorney in the state of Oklahoma, worked as the first female member of the legal department of the magazine, perhaps to change the narrative that the previous article put into the culture of the Mobil company (Magnolia News 1957c).

The publication *On Stream* promoted news from around the work, and portrayed Jaiyeola Aduke Moore, who was a working Nigerian lawyer, and was now the director of Mobil in Nigeria. This news announcement was made in 1957–1958, in hopes of changing the narrative at Mobil. Articles like "Are You a Handicap" likely influenced the company culture, as it was a publication read by employees and marked to employees (On Stream 1957d). Moore would appear later in 1964 in the *Baltimore Afro-American*, a newspaper, in the announcement of her doctorate from Columbia. Moore was also working for the United Nations (Graham 1964). She was honored with a Doctor of Laws by Columbia University, where Moore addressed graduates. She stated that the value of higher education was important in nations like Nigeria, but also relevantly novel in the United States, perhaps to contextualize the challenges that the United States had to face in regards to women accessing higher education, "'higher education for women even in America had been comparatively recent. In a developing country such has my own, Nigeria, the idea of formal education for women is very new'" (Graham 1964; Moore 1966). She expressed optimism as women having access to university education and having the ability to change their lives.

7.9 Oil, Women, and Culture

At the same time that many of these pamphlets and studies about women entering the geology profession were being published, the oil industry tried to make itself attractive to female consumers, and women were thinking about entering the oil profession. The oil industry wanted to attract women as consumers, but was also open to women continuing to joining the profession, and marketing seemed to follow suit. Looking at material culture and literature of the early to mid-twentieth century shows the engagement of the oil industry with women in both consumption, identity, and employment.

The image of the geologist was changing but practicing geologists wanted more representation in science-fiction. In 1961, complaining about the lack of geologists represented in fiction, Mark W. Pangborn, Jr., laments that more books have focused on the exciting exploits of the nuclear scientist (Pangborn Jr. 1961). He hoped that the geologist would also receive more attention. And there is a lot of room for exciting and interesting plots involving geologists. He also includes the idea that there were not any books that focused on the work of female geologists or had a female geologist as a main character:

> The nuclear scientist[24] has been the subject of several worth-while novels in recent years. We can hope that the geologist, too, will receive similar attention from thoughtful writers, whether scientist or professional novelist. Unhackneyed plots are abundant. For example,

> no novels have ever been based on the exciting lives that that our pioneering geologists led in the Old West. No author has ever pitted an honest State geologist against the politics and disappointments that sometimes crop up in State capitals, nor has matched an enthusiastic field man against a neurotic wife who resents his long absences. The problems faced by the female geologist who invaded what is essentially a man's profession must be worthy of a novel; can we ignore the perils that face her in the field? (Pangborn Jr. 1961, pg. 52)

Interestingly, Mark W. Pangborn, Jr., calls for scientists to take up creative writing, though he cautions them from taking up literature. He called for his profession to produce their version of *Arrowsmith. Arrowsmith* was the 1925 novel by Upton Sinclair. The novel glamorized the seeming heroic nature of becoming a physician, doing disease research, and ending epidemics, while at the same time becoming a romantic hero. He promotes the importance of geologist in fiction because it will raise the place of geology in the public consciousness,

> It seems likely that the rising interest in the personality and activities of the scientist will result in more and more stories about our profession. If it is too much to expect another *Arrnowsmith*, with the geologist as protagonist, let us hope that at least some of these future novels will be meaningful to both the scientist and the public, and adequately reflect the thirst for knowledge the critical outlook, and the self-dedication that sets the scientist apart from the average man. (Pangborn Jr. 1961, pg 53)

And though Mark W. Pangborn, Jr. might not have been calling to glamourize the role of geologist, but perhaps to simply promote its importance in the public sphere and among other scientists, he also called for women's stories to be told along with those of male geologists. Female geologists had appeared, though sporadically, in literature prior to the call of Pangborn in the 1960s. There needs to be more scholarship about women in geology in literature. But women did appear in stories about geology prior to Pangborn's call to action in the 1960s.

William MacHarg wrote the story *The Rockhound*, that appeared in the 1921 edition of *Cosmopolitan*. W. H. D. Koerner illustrated the short story, and he was a successful artist of the American West (Hutchinson 1978). His artwork is shown in Figs. 7.10 and 7.11. The 1921 edition of *Cosmopolitan* lacked many of the quizzes, sex advice, and popular culture engagement that readers today would associate with the magazine. The magazine mostly focused on short stories during the early twentieth century, but transitioned into a magazine that focused on the interests of women in the 1960s (Scanion 2009; Landers 2010; Hauser 2017).[5] *The Rockhound* was a short story that focused on the heroic work of Edwin Atwill, a geologist working in the oil fields of Texas during the oil rush as "All Texas was oil-mad; all Texans carried oil stock in their pockets" (MacHarg 1921, pg 59). Edwin Atwill meets Miss Alida Mason, a woman who was working to get her investment back from her dead brother's company, as he had started an oil company whose well did not work out. The female protagonist worked in an office and was treated as too trusting, as she wanted to have her investment shares in the oil company bought back.

Edwin Atwill took Alida to talk to Sam Weld, an experienced oil man in Texas, and Weld treated Mason as incredibly naive and ignorant of business itself: described Alida Mason as childlike for her desire to have her investments returned,

[5] See the website for the history of the publication.

Some of the most romantic stories of our day are coming out of this new Klondike, the Texas oil-fields. William MacHarg *gives the human side of this new "gold" rush—and a love-story, too.*

The Rockhound

Illustrated by

W. H. D. Koerner

IT WAS on a hot midsummer afternoon that the young lady from the North entered the offices of Meeks & Mather, oil geologists, in Wichita Falls, Texas. Because of the hour and the heat, the offices were almost deserted. She wandered uncertainly through the outer rooms and came, finally, to the inner cubby-hole occupied by the youngest of the firm's field-experts, Edwin Atwill.

Atwill, in the khaki and puttees of the field-geologist, was absorbed in a difficult computation. She stood an instant in the door, observing him. His pencil-point broke; he reached to get another and perceived her. His hand remained suspended over the box of pencils.

Atwill laughed happily. "Then you and I start life together owing Meeks seventeen thousand dollars"

Fig. 7.10 Image from *Cosmopolitan* (MacHarg 1921)

> "Buy it back?" Buy it?" he ejaculated. "Great God!" He stopped looking at Atwill as a kitten in order to look at Alida as a creature in a menagerie. "Me dear young lady," he pursued, when rerecovered breath. "[T]hat shows that you don't understand the oil business. The first fact about oil is that no man knows what he's got. To-day's ten-thousand-barrel gusher may be dry next month: to-day's dry hole, when hope's abandoned, blows itself. That's oil. Nobody buys non-producing stock thinking they're buying anything. It ain't buying. It's betting." (MacHarg 1921, pg. 60)

But Mason argued that she and her brother were deceived into investing by a corrupt promoter. Ultimately Weld and Atwill get Mason the upper hand in the formerly failed company in a hostile takeover through a stock scheme, and then find a profitable well, that was not her brother's well. With many adventure stories of the day, Mason and Atwill fall in love during their triumph and marry. But women transition from the heroine in distress that is saved by the geologist, to a type of glamorous female fatale in the 1960s.

Female geologists were featured in the pop culture of cartoons and newspapers of the 1960s. An example cartoon is shown in Fig. 7.12. Cartoons could be seen a mirror to American popular culture (Knaff 2014; Cocca 2016).

Fig. 7.11 Mason portrayed as vulnerable on the right and then advocating for support for her company in the panel on the right (MacHarg 1921)

Fig. 7.12 Steve Roper cartoon that features a geologist (The Racine Journal Times Sunday Bulletin 1962)

The portrayal of the female geologist in *Steve Roper* denotes a different image of a woman involved in adventurous geology stories. Female geologists became the glamourous, active character in cartoons like *Steve Roper*. The cartoon strip went through many changes and many of the plotlines involve American Indian culture.[6] The series involves two men, Mike Nomad, a World War II veteran who had work

[6] Unfortunately, there is little formal scholarship on the cartoon. One of the best sources, but perhaps not as reliable as a traditional academic book or article comes from the Fandom Wiki: https://heykidscomics.fandom.com/wiki/Steve_Roper_and_Mike_Nomad#References

experience in the petroleum industry, and Mike Roper, a photographer and educated character from a well-to-do background, who is not as gruff as Nomad.

The geologist character was noted for her "brains" and "beauty." The geologist is named "Barbie" and worked as a "consultant geologist" for the oil company Hi-Strike. Roper does not take the strange flower delivery seriously and explained to Barbie that she was attractive and should get flowers. The flower sender took on some stalker-like tendencies and might threaten her life. The character Barbie is drawn as being fashionable, beautiful, modern, and being a geologist. The plot line involved Barbie, who was being harassed by a villain named Bert Lorpa. The plot continued where they were entrapped by Barbie pretending to be a princess Barbara. Roper was credited with saving Barbie's life and bringing justice to those persons who were swindled out of their oil investments. I have selected a couple of panels from around May to June of 1962 that illustrate the major points of the story and that highlight Barbie's work as a geologist. Cartoons ran in different stages in different newspapers during the times, so you could have the sequence appearing earlier or later depending on the newspaper. I did my best to stitch them together in a coherent narrative, which appears in Fig. 7.13.

The summary of the narrative is that Princess Swift Arrow, named Barbie by Steve Roper, was a geologist. Mike Nomand knew her father and she contacted him because she believed that her life was in danger. Nomand had worked with her father in the oil fields. Barbie was royalty in a Native American tribe that is not named in North America, as Nomand noted that her father was a "chief" and her mother was a teacher. Bert Lorpa was her "secret admirer" and was stalking her as he was in prison, along with her father and Mike Nomand, for stealing oil profits from the Indians. Barbie, was a small child at the time, and had no knowledge of the events. Lorpa, it is later revealed, had escaped from prison and did not serve his full term. He ultimately coerces Barbie to go out to a dinner date with him. Barbie, with the help of a photographer Chuck Steele, hopes to get a photograph of him and send it to the police. Chuck Stelle fell in love with Barbie, and Nomad was trying to get them together.

Steele has one arm, which the cartoon calls a "handicap." Nomad informs Barbie of Steele's physical appearance of having a single arm in order to determine if Barbie would accept him. Steele is upset and defensive that Nomad would take those actions without informing him. However, Barbie is interested in him, but Steele is offended that she said that she was not bothered by his physical appearance. Ultimately, Steele patches up their relationship, but they are attacked by the escaped Lorpa. But Lorpa is beaten up by Steele and police are called to arrest the escaped convict. However, Lorpa tries to frame Barbie for stealing maps, as Lorpa had millions of dollars in cash on his person. It is revealed that Barbie is innocent, marries Steele, and continues her job working in the oil industry.

Women seemed to have more agency in narratives in the 1960s that were conveyed in cartoons like *Steve Roper*, and were portrayed as glamorous and fashionable. Characters like Barbie continued to have technical and highly scientific jobs, and their geological competence was paired with physical attractiveness. Barbie was also part of a narrative of exoticism and eroticism in the male mind,

Fig. 7.13 A constructive narrative of the Barbie arc in *Steve Roper*. (I looked at cartoons from May 1962 to around late July 1962) (Roanoke Times 1962a, b; Lubbock Avalanche-Journal 1962; Citizens-News 1962)

with the woman, though scientifically competent, remaining a character that needed to be rescued and romanced (Marubbio 2006). This trope would continue into the 1980s with films in the James Bond franchise.

Into the 1980s, the female geologist would continue to be used in movies and popular culture as a female fatale. In 1985, representation of female geologists continued to emphasize the glamorous and "bombshell" in the fourteenth James Bond movie *A View to A Kill*. *A View to A Kill* was a notable film, as the movie featured African American actresses and a storyline dealing with petroleum, involving a female geologist. Stacy Sutton, portrayed by Tanya Roberts, was not reviewed well by fans or critics. Scholars examining women in James Bond films categorize Sutton as an assistant, but not highly involved in the narrative, and as the scholar Stephen Nepa described, "...assist Bond but do not overshadow him as a primary hero."[7] Nepa analogizes to other female scientists in other Bond films like *Moonraker*, and technically savvy women in *License to Kill* simply provide assistance in the completion of a task for Bond, "While these Bond Girls possess intelligence or a skill set that aids Bond on his missions, they also serve as easy targets who put little fuss in sharing Bond's bed, only to be forgotten by the start of the next film" (Nepa 2015, pg. 194). Bond characters like Sutton were associated with "...factiously rolling-back the advances of feminism in order to restore an imaginarily more secure phallocentric conception of gender relations..." (Nepa 2015, pg. 194).

Geologists in the 1950s and 1960s were immersed in American and Western culture, especially ideas about gender. However, female geologists that were profiled in newspapers in the 1970s began commenting on how they responded and personally felt about that culture to a more significant degree than those women in World War II. One example can be found in the profile of J. K. Lentin, an American geologist studying in Canada. She comments on the indignity she had experienced through the assumptions of male co-workers and started to note the "hassles" that she experienced being a woman in Western culture and petroleum geology.

7.10 Female Geologists Comment on the Culture That They Live and Practice In

The ideas and sentiments about women in the 1957 article likely underscored the position and attitudes of the management and company culture regarding women within the company, but the letters coming out against the article also indicate the push back that women in petroleum geology engaged in in the postwar era. There are a lot of historical issues going on in this time period, from the transition of women out of the household and into the working environment, but also the rise of

[7] For a cast listing, please see the film's Internet Movie Database profile: https://www.imdb.com/title/tt0090264/

a second wave feminism empowering women to seek better working conditions, or advocate for their own personal liberties at oil companies, such as dressing how they felt.

There was push back against the empowered geologist. One example can be found in the Canadian newspaper *The Calgary Herald*, where J. K. Lentin was profiled in 1974. The article was entitled "Palynologist's path to independence paved with…" and the subtitle continued with "…Dinoflagellates, beer and dirty jokes" (Gilchrist 1974). The article was written by a female journalist and Gilchrist wrote about the surprising presence of Lentin, who went to the Oilfield Technical Society in Halifax. Gilchrist wrote that most of the audience thought that Lentin was simply there to "entertain" the members as she was a woman, and this was Lentin's last meeting as "She didn't attend any others" and now Lentin "…laughs about that meeting now" (Gilchrist 1974). The attendees of the dinner thought that she was an exotic dancer, or "stripper." The narrative of the article shifted to Lentin's position that she could participate in petroleum geology, but did not seek to cause "hassles" (Gilchrist 1974).

But these hassles that Lentin was trying to avoid were brought about by working in oil as a woman. And Gilchrist continued the narrative of women working in petroleum as being exceptional: "And in the man's world of the petroleum industry, it's very easy for a woman to create hassles, just by her very presence" (Gilchrist 1974). Lentin was a scientist who studied fossilized pollen, and the small animal life (dinoflagellates) that often appear in fossilized pollen.

Lentin was working as a scientist who was not tied down to a single oil company. She wanted to become a professor, but left her job at an oil company to become an independent consultant. She was confident about working on her own, "I'm not even slightly frightened at the prospects. As a matter of fact I'm looking forward to it with enormous joy" (Gilchrist 1974). And she asserted that she was just as capable as a man working as a consultant, but men who worked on their own were treated with more respect. She compared the experiences of men and women in consulting: "'An independent man can get along. They call him an eccentric. ["] But she added that, "An independent woman they call a bitch" (Gilchrist 1974).

She rose through the oil industry, determining that a woman could not rise higher than entry level with only a bachelor's degree, and decided to enter graduate school studying fossilized pollen because she was told that was an important need for oil companies. She had to pay for her own graduate education, as the grants to pursue higher education was only going to men in the company. And she got her doctorate in only two years, and she paid her own way.

Even though she held a doctorate when Lentin returned to the United States, industry did not value her skills, nor her expertise in palynology, as there was a market saturation of palynologists in the oil industry. At one job interview, she was treated as if she was interviewing to become a secretary: "In one job interview, as she sat there with her Ph. D. thesis tucked under her arm, she was asked "How many words a minute do you type" (Gilchrist 1974). In 1974, Lentin was told explicitly that she would not be hired because of her sex, "In another company, after a series of interviews that fully established her credentials and her qualifications for the job, the senior geologist told her bluntly he would not hire her because he would not have a woman working in his department" (Gilchrist 1974). Lentin referred him to the changes in law which encouraged more equitable hiring practices: "But that's against the law," she said in amazement. "So, sue me," he replied" (Gilchrist 1974).

Lentin was hit with a challenge that was perhaps more subtly implied for women working in geology, but now was a vivid reality in the 1970s. Though Lentin could take her potential employer to court and possibly win, she would worry that there would be a cultural and social retaliation at her job. Lentin was enmeshed in dilemmas that challenged women in the 1970s until today, though the law was technically on her side, culture, politics, and society within the industry could be used in retaliation, "She [Lentin] knew she could not win that fight. If she won in court, she'd have her job but life in that department would be made miserable for her./And word of any court challenge could spread around the industry and other jobs would be closed to her./ 'You just don't sue the oil industry,' she said" (Gilchrist 1974).

Lentin was effectively driven out of industry, and this example represents a change in reporting with female petroleum geologists. Previously, female petroleum geologists at the turn of the twentieth century and during the wars went with the image the oil industry promoted: the oil industry was open for business for women, and though they were exceptional, oil and gas was an exciting place for women, a business to aspire to become a part of in their long-term career plans.

The 1970s has several examples of breaks in messages where discrimination and cultural and social retribution was a real experience that was acknowledged by some female petroleum geologists like Lentin. Lentin was effectively driven from the field, and had to work as a "head hunter," who worked to fill employment positions for companies. But she returned to the oil industry around two years later in Canada. She then discussed the exceptional number of paleontologists who worked in North America in general, and could count them on both hands, including those in the United Kingdom. Her image is shown in Fig. 7.14. Russia, she added, had female paleologists "…all over the place" (Gilchrist 1974).

Lentin, in a previously completed interview, mentioned that there was a very small number of women working in paleontology in the United States and Britain. Women working in paleonotlogy, according to Lentin, were in much higher numbers. The United States, Lentin implied, and unlike Russia, likley socialized people into the expectation that women did not take jobs working in paleontology. She commented on the social life and expectations regarding women and men in the United States, "It all comes down to what people are conditioned to, she said, and home conditioning, until the last few years, was that women were in the home washing dishes, scrubbing floors and soothing children when they have the sniffles" (Gilchrist 1974). Men in the United States still associated domestic work with women, like washing dishes. And it was a cognitive leap because of how society treated men and women, for men to accept equity with women, "It is difficult for a man to accept a woman as an intellectual equal, or an intellectual superior—especially an intellectual superior" (Gilchrist 1974).

But she also emphasized and drew a dichotomy for women in the field of geology as well. And Lentin departed from the dichotomy of glamor and doing dirty field work, and put more emphasis on the dirtiness and grit of fieldwork. She

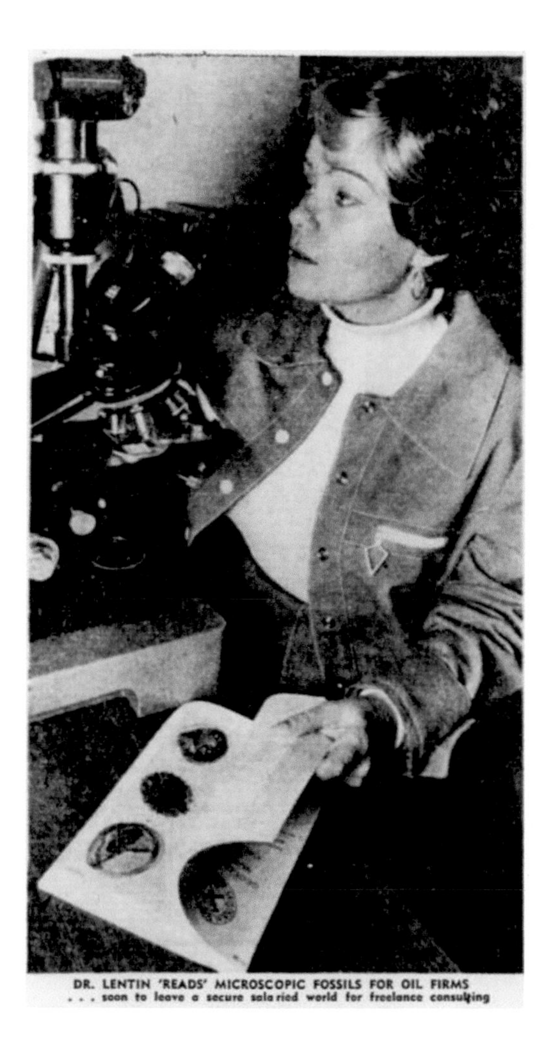

Fig. 7.14 Dr. Lentin at work in the lab (Gilchrist 1974)

cautioned that, "Geology is not place for someone who is prim and proper, and according to Dr. Lentin[,] Even at the undergraduate level, a woman must-be able to handle a class situation where most classmates and instructors are male" (Gilchrist 1974). Women had to be tough and deal with a dangerous field of work: "'She has to be able to cope with very rough talk, often talk that's put on just because she is there" (Gilchrist 1974). Though Lentin does not explicate what the talk consisted, an assumption of hyper masculinity or misogyny probably is the subject of the warning.

The field was dangerous, and trips required in geological education were also dangerous, but these trips were dangerous for anyone, including men. She warned of fields where she "…recall[ed] trips into scorpion and rattlesnake country where men, too, had to watch their step" (Gilchrist 1974). But in harkening back to the narratives of June King, she has to put in with that same masculine culture: "She has to be able to sit around the fire, drink beer with the boys, tell dirty jokes and laugh at dirty jokes" (Gilchrist 1974).

Clothes, again, become something of both a personal and professional identity for female geologist working in oil. She had similar criticisms as found in the work of Mary Fiedorek in *Executive Style*, which was written around the same time that Lentin is profiled in the *The Calgary Herald*. Lentin criticized female geologists from becoming too masculine, "Unfortunately these conditions [the field] have led some young women into sloppy and masculine dress and habits, she said" (Gilchrist 1974). And according to Lentin, most of the petroleum industry would not hire a woman who dressed too masculine, "Most oil companies reject this kind of woman. It's bad for their image. Women should look like women" (Gilchrist 1974).

Nonetheless, she was striking out on her own, as she was optimistic that the oil industry, while cutting back jobs in the 1970s, would hire a consultant like her, rather than hiring on more employees for oil well analysis work. Lentin also wanted to return to university life, among other things, "...because she wants to be there to encourage young women to become scientists" (Gilchrist 1974). She wanted to defy the convention wisdom of the time that she needed to work at a lower reputation university and then seek employment at a more prestigious institution later, and wanted to "...tackle situations head on [,]" as that was her "style" (Gilchrist 1974).

Lentin further thought about her place in a changing scientific landscape where women had more opportunities, and was more critical, and less praising of both science and the oil industry. She argued that in a generation, or maybe more time, things would be different for women in science: "'It will take little boys whose mothers are doctors, lawyers and Indian chiefs for them to grow into men who can cope with women being doctors, lawyers and Indian chiefs" (Gilchrist 1974).

In the next chapter, women will continue to voice their criticisms and comment on their experiences in science. These comments increased because of the acceleration of the women's movement and efforts like the Equal Rights Amendment. Women in newspapers would still continue to be portrayed as glamorous working with geology, but found more room to comment on their experiences in work and science.

References

(1943) War living: aprons. Life, 22 February, pp 66–68

(1944) From Dimity to Denim. Petrol Eng 15(13):134–136

(1945) Women in Oil Jobs: Strike Oil. Glamour. July. Briscoe Center for American History. Series I, Subject Files, 1912–2000. Exxon Mobile Historical Collection, unspecified — Box: 2.207/L13A in Exxon Mobil. (1909–2000). Reprints [External Magazine Articles]. Corporate, 1942–1999 in Exon Mobil Historical Collection

(1953) Women at work. Petrol Eng 25(13):E-1–E-2

(1954a) Women at work. Petrol Eng 26(1):E-36–E-37

(1954b) Women at work. Petrol Eng 26(2):E-23–E-26

(1955) Women at work. Petrol Eng 27(1):E-11

(1957a) Are you a handicap to your husband? The Magnolia News. July–August. Found in Biscoe Center for American History at the University of Texas at Austin. Exxon Mobil Historical Collection: Corporate Administration, Human Resources, Women, 1917–1985
(1957b) Who's a handicap you bum. The Magnolia News. July–August. Found in Biscoe Center for American History at the University of Texas at Austin. Exxon Mobil Historical Collection: Corporate Administration, Human Resources, Women, 1917–1985
(1957c) Mary Elizabeth Lewis Added to Magnolia Legal Staff. The Magnolia NewsDecember: pg. 8. Found in Biscoe Center for American History at the University of Texas at Austin. Exxon Mobil Historical Collection: Corporate Administration, Human Resources, Women, 1917–1985
(1957d) On stream: woman director. Found in Biscoe Center for American History at the University of Texas at Austin. Exxon Mobil Historical Collection: Corporate Administration, Human Resources, Women, 1917–1985
(1959) Industrial contest. The Leader-Post, 3 February
(1962) Steve roper. The Racine Journal-Times Sunday Bulletin, 6 May
(1962a) Steve roper. Roanoke Times, 5 May
(1962b). Steve roper. Roanoke Times, 8 May
(1962) Steve roper. Lubbock-Avalanche-Journal, 3 June
(1962) Steve roper. Citizen-News, 27 July
(1970) Legs vanish as pant-suits invade Canadian offices. The Calgary Herald, 3 October
(1971) "A Rendezous with destiny": three wives of Apollo 14. Bangor Daily News 26 January
(1972) Uniform sale. The Tuscaloosa News. 7 May
(1974) Crude and refined. Oilweek 25(13):28
(1975) Effective communication for engineers. A collection of articles which were originally published in chemical engineering. McGraw Hill, New Yor
Anthony CS (2023) Camera girl: the coming of age of Jackie Bouvier Kennedy. Gallery Books, New York
Baker CN (2004) Race, class, and sexual harassment in the 1970s. Fem Stud 30(1):7–27
Buszek ME (2006) Pin-up grrrls: feminism, sexuality, popular culture. Duke University Press, Durham
Coll S (2013) Private empire: exxonmobil and american power. Penguin, New York
Chrisman-Campbell K (2022) Skirts: fashioning modern femininity in the twentieth century. St. Martin's Press, New York
Cocca C (2016) Superwomen: gender, power, and representation. New York: Bloomsbury.
Doan C (1961) Beauty consultant gives sage advice. The Leader-Post. 21 April.
Elias A (2022) The rise of corporate feminism: women in the american office, 1960–1990. Columbia University Press, New York
Exxon Mobil (1917–1985) Human Resource Records: Women 1917-1985. Box 2.207/E46. ExxonMobil Historical Collection. ExxonMobil Historical Collection, 1872-2003. Briscoe Center for American History. Scanned by Briscoe Center
Friedorek MB, Jewell DL (1983) Executive style: looking it, living it. New Century Publishers, Inc
Gilchrist M (1974) Palynologist's path to independence paved with…dinoflagellates, beer and dirty jokes. The Calgary Herald, 12 December
Golding M, Stitles, R (2022) Adams, Ernestine. Handbook of Texas. Online: https://www.tshaonline.org/handbook/entries/adams-ernestine
Graham GP (1964) Brilliant african lawyer gets colubmia doctorate. Baltimore-Afro-American, 24 November
Green V (1952) Woman's place in management. Aman Gas Association Monthly 34(11):11–12, then continued on pg. 37
Harkess S (1985) Women's occupational experiences in the 1970s: sociology and economics. Signs 10(3):495–516
Harrington C (2020) What is "toxic masculinity" and why does it matter? Men Masculinities 24(2):345–352

Hauser B (2017) Enter helen: the invention of helen gurley brown and the rise of the modern single woman. Harper, New York

Hillman BL (2015) Dressing for the culture wars: style and the politics of self-perception in the 1960s and 1970s. University of Nebraska Press, Lincoln

Hutchinson WH (1978) The world, the work, and the west of W. H. D. Koerner. University of Oklahoma Press, Norman

Jackson LM (2023) The invention of "the male gaze." The New Yorker, 14 July. https://www.newyorker.com/books/second-read/the-invention-of-the-male-gaze?source=Paid:Soc_FBIG_CM_0_DPA_GTM_0_NYR_US_Prospecting_C&utm_source=facebook&utm_medium=Paid:Soc_FBIG_CM&utm_brand=tny&utm_campaign=paid-DPA_GTM&fbclid=IwAR332KmdkbsJgZg7_8xBWx9Xu4uWRwdLvAR-Bb5s8qWyZdPE5Lj8uXzuIEY

Johnson DK (2004) The lavender scare: the cold war persecution of gays and lesbians in the federal government. University of Chicago Press, Chicago

Knaff DB (2014) Beyond roise the riveter: women of world war ii in american popular graphic art. Lawrence: University of Kansas Press.

Landers J (2010) The improbable first century of cosmopolitan magazine. University of Missouri Press, Columbia

Levy A (2006) Female chauvinist pigs: women and the rise of raunch culture. Free Press, New York

MacHarg W (1921) The rockhound. Cosmopolitan 70:58–64, continued towards the end of the volume with unnamed page number

MacKinnon CA (1979) Sexual harassment of working women: a case of sex discrimination. Yale University Press, New Haven

Marubbio EM (2006) Killing the Indian maiden: images of native american women in film. The University Press of Kentucky, Lexington

Matelski M (2017) Reducing bodies: mass culture and the female figure in postwar america. New York: Routledge.

Moore JA (1966) Struggle for political unity in nigeria: lecture, auspices of department of political science vassar college. Scan from Vassar Archives. Pamphlet DT515.8 .M66 1966

Nepa S (2015) Marriage, gender roles, and the "different bond woman" in *on her magesty's secret service*. In: Funnel L (ed) For his eyes only: the women of james bond. Columbia University Press, New York

Osgerby B (2001) Playboys in paradise: masculinity, youth and leisure-style in modern america. Berg Publishers, Oxford

Pangborn MW Jr (1961) Geology and geologists in fiction. J Wash Acad Sci 51(4):49–53

Putnam E (1983) Executive women now sport feminine attire. Ottawa Citizen, 10 May

Reichert T (2003) The erotic history of advertising. Prometheus Books, Amherst, New York

Robb I (1953) Oil and insurance fields wide open for women. St Joseph News-Press, 15 May

Scherer R (1982) Short skirt is not yet labeled taboo at ibm. Boca Raton News, 1 August

Scanion J (2009) New York: bad girls go everywhere: the life of Helen Gurley Brown. Penguin Books

Shelmire RW (1919) The draftsman. Scientific Publishing Bureau, Chicago

Stratas D (1967) Mod clothes out for job hunters. The Calgary Herald, 23 August

United States Congress (1951) Congressional record: proceedings and debates, volume 97, part 12. United States Government Printing Office, Washington, DC, p A2725

Wdowik M (2017) The long, strange history of dieting fads. The Conversation. 6 November.

Whitt J (2008) Women in American journalism: a new history. Univerisity of Illinois Press, Champaign

Women's Focus Group (1951–1968/1962–1963) Women's Discussion Group. Box 2.207/E34. ExxonMobil Historical Collection, 1872-2003. Briscoe Center for American History. Scanned by Briscoe Center

Chapter 8
Identities, Politics, and Culture in Flux: Bodily Autonomy, Political Rights, the ERA, and Breaking from Established Narratives

8.1 Introduction: Revolutions

The late 1960s, 1970s, and 1980s brought about radical challenges to the prescribed image of the glamorous petroleum geologist instituted in the newspapers and industry from the early 1920s into World War II, and then into the postwar era. The challenges of the Equal Rights Amendment (ERA), sexual revolution, and calls for racial equality work themselves through the print culture of the twentieth century. Female geologists continued to be portrayed by the petroleum industry as the exceptional, but acceptable workers in geology and other jobs related to petroleum. However, some women began to criticize and defy conventional standards for what people imagined as female petroleum geologist or challenge the idea of women in the sciences. The chapter explores the changing and challenging landscape for women in petroleum and the sciences, and how some female students, workers, and scientists experienced the Sexual Revolution of the 1970s and how they both rebelled and embraced the oil industry.

8.2 Female Geologists and *Playboy Magazine*

As this book has discussed in previous chapters, women were sexualized in their jobs as geologists in World War II and beyond. The physical appearances of some of these female geologists were described with a plethora of adjectives that would be unprofessional at best in today's scientific landscape. In the case of June King, her weight was described, and again commenting on someone's weight in today's society is both socially rude, unprofessional, and frankly none of anyone's business, not to mention the problems with fatphobic language in current United States Culture (Matelski 2017; Wdowik 2017).

E. A. Driggers, *Glamour and Geology*,
https://doi.org/10.1007/978-3-031-64525-9_8

Aspects of pin-up culture and the acceptance of pornography in the 1970s likely led to some college students and college educated women appearing in the pornographic magazine *Playboy* (Bronstein and Strub 2016; Fraterrigo 2009). *Playboy*, for historians, represents not only many historical insights but also many problems. Historicizing *Playboy* magazine is difficult for several reasons. The debate about whether *Playboy* is exploitative or empowering of women is a problem sufficient for a book length project, even setting aside any definitions of pornography. As a historian, I feel that pornography is exploitative, as many of the women who posed for men's magazines like *Playboy* were not able to profit off their likeness and images in the same manner that corporations like *Playboy* were able to generate large amounts of wealth (Dines et al. 1998).

Putting those issues aside for other books and projects, I would like to focus our discussion in the historical place of *Playboy* in the women's rights movement and its meaning in history. Scholars of sexual and gender politics, like Carrie Pitzulo, have placed *Playboy* in a broader context of meaning in the political history of the 1970s. Pitzulo has given *Playboy* an evenhanded treatment in terms of gender and political history, but has largely ignored the economic questions related to likeness and profits.[1] But in terms of assessing *Playboy* in this study, I think her analysis is useful in understanding why an undergraduate geology major would pose for the magazine and what that meant for women in geology and society.

Ptitzulo has contextualized in pieces like "The Battle in Every Man's Bed: Playboy and the Fiery Feminists," that there was a lot of tension in the 1970s women's movement and women's feelings about *Playboy* magazine (Pitzulo 2008). Prior to the 1970s, Gloria Stenium, a journalist, went undercover into the *Playboy*-owned clubs, where servers had to dress in tight and restrictive bunny costumes and serve drinks to patrons who were often rude and sexually aggressive. The Steinem expose put *Playboy* on notice with feminists and resonated with angry frustration by some feminists in the movement. The 1970s feminist movement was not monolithic, and included those who were "Sex positive" or women who argued for the progressive and empowering effects of pornography and those women who were not in support of pornography (Pitzulo 2008). Some feminist critiques of magazines like *Playboy* centered on the problem of duality of women: they could either be a virtuous "Madonna" (such as the Virgin Mary in Christianity) or a "whore." Feminists often critiqued both society and the *Playboy* magazine, citing this problematic duality. Criticism was pointed at the magazine by some feminist historians like Pitzulo who argued that the magazine itself played some role in progressive politics of the 1970s and supporting specific parts of the women's movement: "a progressive stance on women's rights through the 1960s and 1970s and was particularly vocal in support of abortion" (Pitzulo 2008). But continued criticism of the magazine centered on the "exportation," "objectification," and "degradation" of the women who posed for

[1] Please see her full-length book project (Pitzulo 2011). The question of profiting off of likenesses is a question in many fields, including NCAA athletics and exploitation.

Playboy (Fraterrigo 2009).[2] It was also deemed a space for "casual misogyny." (Harrington 2020 and Pitzulo 2008)

One geologist who posed for *Playboy* was Martha Thomsen. Thomsen is being discussed here because she continues the sexualization and feminization of women in petroleum geology, but she discusses her ideas about bodily autonomy, as well as her views about posing for a nude magazine. Unfortunately, June King and many of the other women profiled in this book are not on record for their feelings about participating in the advertising for oil companies. It is unclear if any of those women also profited from their likeness and images. However, the pin-up culture images were appropriate enough to publish in newspapers, so critics could bring that point to the analysis.

Though historians have offered analysis that focused on the sexism and anti-feminist aspects of the magazine, Pitzulo argued that there was more to the magazine than "naked women." She argued that the, "The magazine hosted important discussion about women's liberation. Nonetheless, there has been little consideration of the way in which, apart from its centerfolds, Playboy's words and money may have contributed to the growth of feminism. Indeed, by the early 1970s the magazine served as a regular, progressive, and mainstream forum for discussions of expanding roles in society" (Pitzulo 2008, pg. 260). And she argued that the magazine supported Civil Rights in the 1960s and then had to support the later women's movement, but he supported a specific version of the women's movement,

> By 1970, however, Hefner perceived two versions of feminism. One was supposedly rational and mainstream faction that promoted liberal goals like antidiscrimination laws, and the other was an extreme and militant version that allegedly wanted to overturn heterosexuality. Hefner's simply dichotomy failed to comprehend the diversity of the feminist movement, which included not only these two poles but also a wide range of activist that combined varying goals, agendas, and priorities. (Pitzulo 2008)

It must be acknowledged that the progressive branding of *Playboy* has been injured by the legacies of its founding editor and his associates (Bronstein and Strub 2016; Fraterrigo 2009; Pitzulo 2008, 2011).

Neither of the women discussed in this book actually became working geologists. Both women had different careers, as Thomsen worked in entertainment and at one point was a stewardess. Nonetheless, discussing these two women provide a limited, but partial glimpse into how women were viewing their bodies, actions, and rights associates (Bronstein and Strub 2016; Fraterrigo 2009; Osgerby 2001; Reichert 2003; Pitzulo 2008, 2011). Thomsen was also represented, though likely through the marketing and public relations department of Playboy, as a "former geology major."

Heterosexual men and women differed at the time about their attitudes regarding sexual material. Unfortunately, most of the research performed at the time did not expand itself to the variety of sexual orientations, such as bisexual or homosexual,

[2] There is a robust debate that continues about the nature of pornography and its exploitation of women. See Dines 2010 for an example of an anti-pornography feminist.

and were often framed in heteronormativity (Katz 2007). In 1976, the *Williamson Daily News* ran an Associated Press story that a, "Survey suggests women are still affected by traditional pressures. Researchers in Philadelphia examined nude images of men and women in magazines to gage people's attitudes regarding sex. The survey was conducted with 100 students from local colleges and half of the participants were women and the other half were men. The images were taken from pictorials from *Playboy* and *Playgirl* (a pornographic magazine which featured nude men).[3] One third of the women reported that examining the images made them feel "guilty." The researchers reported that women wanted more authentic sexuality, "Most women expressed a desire for a more realistic, contextual representation of sexuality—one in which a relationship was clearly defined between the participants and could reasonably lead to nudity and sexual activity[.]" The men surveyed were the opposite, "In contrast, 75 per cent of the men rated the nude centerfold of a men's magazine sexually stimulating, the two said, and 84 per cent said they would buy the magazine."

Thomsen was interviewed by Bill Sallquist in the *Spokane Daily Chronicle* in 1980. The article opens with the most provocative descriptions meant to entice the reader to continue the column, "Martha Thomsen, the Washington born beauty who blushes at having to disrobe for a physical, is the same Martha Thomsen who smiles back at you from the center pages of the May issue of Playboy Magazine" (Sallquist 1980). Her biography is communicated in the article, she was twenty-three in 1980 and was formerly featured in a pictorial of students who studied in the colleges associated with the Pac-10 Athletic Conference in 1978, of which Washington State University was a member. She was identified as a "former geology major at Washington State University…" (Sallquist 1980).

Thomsen was not the only Playmate model who expressed an interest in studying geology in higher education. Kymberly Herrin, formerly Kimberly Herrin, had posed for *Playboy* in 1981. The *Spokane Daily Chronicle* covered her experiences posing for the magazine. She was described as a former student who attended Santa Barbara City College. In presenting her background she was described as "…interested in geology and botany" (Huessy 1981). She found that posing for *Playboy* was empowering to women, "'I never had any fantasies about being a Playboy centerfold,' she noted. 'And I do think that it (Playboy magazine) shows how wholesome women are—it shows them as superwomen,' she added" (Huessy 1981). Herrin continued to emphasize how posing for the magazine made women powerful. The article included some additional quotes from Herrin, "The 'superwoman,' not the 'exploited woman,' image is reflected in Kymberly's and Playboy's interest in the body beautiful and a healthful outlook on sports, she said"[4] (Huessy 1981).

Herrin, whose image is given in Fig. 8.1, was described in the article as having an interest in geology, was also described as an adventurer. The article described her

[3] There is not much scholarship on the history of *Playgirl* compared with that of *Playboy*.

[4] The article had some damage to the print on the right side of the column, and those quotes are recorded as best as possible.

Fig. 8.1 Kymberly Herrin. (Huessy 1981)

glamorous lifestyle, not unlike the description of other geologists in this book, "She's a surfer, a scuba diver, and enjoys motorcycling and skiing. Adventure is the byword in her choice of sports. It's a 'thrill' lifestyle" (Huessy 1981). She also described her experiences as giving her more self-confidence, but Herrin made light of the stalking that came from "The glamour side of that lifestyle…" She described men following her home in cars, but she downplayed it, by sharing a story of her following men home in high school.

Sallquish described Thomsen's physical appearance at the beginning of the interview, "Miss Thomsen, who arrived for a Chronicle interview tastefully attired in a tailored beige pantsuit, said, 'I haven't always thought of myself as Playmate material'" (Sallquist 1980). Her road to posing as a monthly featured model started when a representative from the magazine approached her as a student. She said no at first, as she felt nude poising did not have enough monetary or professional advantages. But the magazine continued to appear to hear after her pictures appeared in 1978, presumably that were non-nude. And then she was offered a shot as a monthly model, and she commented on her now found appreciation for her body, "'I guess they liked my smile,' the blue-eyed, honey-blonde Miss Thomsen said, demonstrating a heart-melting grin, 'There wasn't much else showing'" (Sallquist 1980). After presumably completing her studies, she was at work as a flight stewardess for Eastern Airlines.

Working for airlines was still a glamorous position for women in the 1980s, though not as glamourous has it had once been in the days of Pan-America (Cooke 2021; Wohl 2015; Vantoch 2013; Levine 2007). She had applied for the job "on a lark" (Sallquist 1980). Pan-American Airlines, and the jobs of stewardess, was seen as an ultra-glamorous and ultra-feminine job during the 1960s, 1970s, and early 1980s. Thomsen's image appears in Fig. 8.2. The Pan-American airlines culture and times have been routinely romanticized in modern pop-culture, such as in the 2002 movie *Catch Me if You Can* and the American Broadcasting Channel's short-lived 2011 show *Pan-Am*.

Though she had a background in pageants, she was mostly supported by friends and family. And her friends and family helped encouraged her to take the pictorial work. The article argued that the pictorial had been viewed mostly in a positive light. The notable exception was an unnamed Tecoma "broadcaster" who had questioned why she had taken up the nude modeling. She cited her grandma as approving of her magazine work. The article portrayed the nude modeling as an important part of women's rights. *Playboy* received criticism from women's rights activists during the 1970s and 1980s, and *Playboy* and its editorial chief and daughter, who

Washington-born Playmate Martha Thomsen: "It's something, I think, that most girls fantasize about."

A Beauty Bares Views on Nudity

Fig. 8.2 Martha Thomsen. (Sallquist 1980)

was now CEO of Playboy Enterprises pushed back at *Playboy* being anti-feminist (Levine 2007; Kutulas 2017; Pitzulo 2008, 2011).

Thomsen, according to the article, saw her modeling in feminists and empowering terms. Thomsen identified herself as someone who supported women's rights: "'I've always been strongly for women's rights,' she said, adding that her support for the Equal Rights amendment is unswerving" (Sallquist 1980; Fraterrigo 2009). She also defended her right to model nude. She cast the nude pictorial in positive terms: "'The nudity itself is not degrading,' Miss Thomsen said. 'People enjoy looking at something that's pleasing to them'" (Sallquist 1980). And she said that she was appreciative of the new fame the pictorial brought her: "Of the attention it's gotten her, she said simply, 'I like it. I think I'd be very offended if nobody found me attractive'" (Sallquist 1980). She has received attention from customers, as well as minor male customers. She also defended her physical appearance as in keeping with the "Girl Next Door" theme as she said her neighbors would agree. The article then ended with her plans, and she said they were not specific, and she aspired to continue her life, as the most positive events in her life have been "spontaneous" (Sallquist 1980). She also responded to additional questions: she had a relationship, did not lose her job because of her pictorial, did not plan to pose for any other magazines that were not "respected" like *Playboy*, and would recommend the experience of posing to other women, and did not follow a diet.

Though the article makes mention of Thomsen's relationship status of having a "semisteady boyfriend," it does nothing to discuss her marriage prospects, household, or domestic roles, or discuss her relationship in depth. This is a change from the June King articles and pin-up culture that began around the 1940s (West 2020). There were some articles that continued to emphasize the most important role of a female geologist to be that of a homemaker, such as the report of the interview of a woman geologist appearing on the radio, who worked as a geologist, prospector, and rodeo champion. And the newspaper article advertised that, "An artist of note, Mrs. Reese will confine most of her radio words to the role of women in geology and how best they can combine a career with the all important one of housewife"(Prescott Evening Courier 1950). The notion of housewife started to fall away in the 1970s with professional write ups in some cases.

8.3 Women Geologists and the Equal Rights Amendment (ERA) and "Woman Power"

Women in geology and the petroleum business were continuing to think about their role in the women's rights movement and discussed their thoughts about the Equal Rights Amendment. The Equal Rights Amendment (ERA) was a proposed amendment to the United States Constitution. Constitutional amendments require two thirds of states voting to approve the amendment for it to be accepted, in addition to both houses of Congress supporting the amendment. Constitutional amendments are rare, with the latest amendment at the time of publication occurring in 1992, with the twenty seventh amendment requiring Congressional pay-raises to be in effect in the next Congress after they are approved (Maxwell and Todd 2018; Spruill 2017; Cobble 2003). The ERA was approved in the House of Representatives in 1970 and 1971, and the Senate approved it in 1972, which caused a vote from the states in a Constitutional convention in each state, where the amendment did not get enough states to approve it (Maxwell and Todd 2018; Spruill 2017; Cobble 2003). There were attempts in the 1980s to call another convention to get the amendment passed, but they were not successful. There are still supporters of the ERA working to call another vote.[5] Ratification of the ERA did not happen because of well-organized efforts by women like Phyllis Schlafly, who argued that the amendment would end protected status of women, such as different labor laws or protections for women (Maxwell and Todd 2018; Spruill 2017; Cobble 2003).

Women working in geology engaged with the ERA. One example of a geologist who left science to engage in the political rights movements of the 1970s was Genevieve Atwood. Atwood is currently professor emeritus at the University of Utah. Her search focused on the geology of Central America and the Great Salt Lake, and she received her M. A. 1973 from Wesleyan University and Ph. D. from the University of Utah in 2006. After working in public policy, she received a Masters of Publica Administration from the University of Utah in 1991. She also had career experiences founding a non-profit education organization that provides outreach for adults and children to improve geology education. She was a candidate for the U.S. House of Representatives for the 1990 election cycle, where she was not elected, and she served six years in the Utah House of Representatives from 1974 to 1981 (Pedersen 1975). Her professional work included uranium studies and hydrology for companies like Ford, Bacon & Davis Utah where she served as a Senior Geologist.[6]

In 1975, Atwood was profiled in an article in *Deseret News* about the rise of female power in politics. The article entitled "'Women Power' has become mighty force in politics" (Pedersen 1975). Atwood is listed as a "28-year-old newcomer" to politics as she was completing her first year in the Utah House. The article recorded her idea that her scientific and geological knowledge would be useful in politics. Like many young people in her generation, she got interested in politics because of the fallout from the Watergate Scandal. The Watergate Scandal occurred at the Watergate Hotel in Washington, D.C., where members of the "Committee to Re-Elect the President," affiliated with the Richard Nixon administration, broke into the rival Democratic party headquarters (Pedersen 1975; Graff 2022). The Scandal resulted in jail time for the Nixon operatives and ultimately Nixon resigning as President (Graff 2022). Atwood saw herself as only having two choices about the future of her political participation: "She felt she either had to remove herself from

[5] For instance, see https://www.equalrightsamendment.org/

[6] Atwood's biographical information was sources from her posted CV on the department of Geology's website at the University of Utah: https://faculty.utah.edu/bytes/curriculumVitae.hml?id=u0074508

the American scene entirely or get deeply involved in government and do her part to make things better" (Pedersen 1975). And she was an advocate for women entering politics: "'I'm glad we have more women in politics. I think a fresh new voice is desperately needed to rejuvenate the system…It's a real ego trip when you win an election. But it's a big responsibility, too'" (Pedersen 1975).

Atwood also embraced the popular exercise of jogging. In 1972, women were able to run the Boston Marathon for the first time, though one woman, Roberta Gibb, ran the marathon without the consent of the organizers in 1966 (Clerici 2014; Tumin 2022; Switzer 2017). Women were able to run the Olympic Marathon, however, in 1984 (Clerici 2014; Tumin 2022; Switzer 2017). The participation of women in sports was a women's rights issue, and with the passing of Title IX of the Education Amendment in 1972 began opening doors for greater participation of women in sports (Melnick 2018).[7] Jogging was also trending in America at this time as well (Gotass 2012; Sears 2015).

Though it seems like simply a hobby or exercise of a personal individual, putting the article in context shows Atwood's actions as a political move as well. The article "Running is way of life for Rep. Atwood" had a double meaning, as it was part of her identity as well. By 1979, Atwood had served three terms in the Utah House and the article portrayed her exercising as a form of celebration and accomplishment to her highly active life. The reporter, George Ferguson, framed her actions as "Genevieve Atwood says she would rather run than walk" (Ferguson 1979).

At the time of the article, Atwood was running three miles a day, and she was portrayed with the type of glamor associated with June King and other World War II era geologists. Her energy and accomplishment where highlighted: "With the 43rd Legislature in session, this vivacious energetic woman half trots from assignment to assignment exuding enthusiasm, confidence, and interest" (Ferguson 1979). Atwood was even portrayed as having "an infectious giggle" (Ferguson 1979). And her abilities appeared in a humbled context, much like those of female geologists in World War II, "I'm a classic example of how someone who was never involved in politics can get involved. I had a non-political background I came from a low-keyed political family" (Ferguson 1979). The article continued much like those of the write ups of petroleum geologists and their work prospecting for oil, as Atwood was working for oil companies in Honduras. Her work, after her academic achievement, was also portrayed in humble terms, "During a year in Europe she did a bit of everything, from being a Paris art gallery secretary to a strawberry packer in Germany and a stablehand in Ireland" (Ferguson 1979).

She used her history of science degree, in addition to her geology Master's Degree to work as a researcher in coal for Bacon and Davis. But again, like the geologists of the 1940s, she was portrayed as a hard worker. Working thirty hours at her job, and then thirty to forty hours per week in the Utah House. Though she had earned a master's degree in geology from Wesylan, she was studying for a master's degree in engineering geology from the University of Utah. She was completing

[7] See https://www.womenssportsfoundation.org/advocacy/history-of-title-ix/

graduate geology coursework while balancing work and politics: "'Geology is such a rapidly changing science, I like to keep up with it by taking at least one class per quarter,' she said" (Ferguson 1979). Her physicality was discussed in the article as well but it was combined with her assertiveness as a politician, "A 5-foot-6 inch brunette with very decisive mannerism and an animated voice, Rep. Atwood considered herself as 'independent voter'."

Turning back to the world of Augusta Kemp, Atwood was portrayed as a woman of accomplishment, but a new type of accomplishment: athletics. Physical domination and sports achievement were also glamourous for women in the era of Title IX, "Keeping in physical shape is a way of life for Rep Atwood. She was president of the athletic association at Bryn Mawr. At the time, she was a nationally ranked tennis player in women's 20-under singles. She played varsity field hockey, was captain of the tennis team and played varsity badminton and junior varsity basketball. She holds a white belt in judo" (Ferguson 1979). The association of glamor with high achievement and extracurricular achievement continues to this day in most of the lives of students in United States secondary schools and universities. In the continuation of achievement, she also became the number one rated women's squash player in the whole state of Utah. Atwood was profiled as a runner in the article, and that image appears in Fig. 8.3.

Atwood was interested in women's participation in politics, as she cited the declining number of women's participation in the Utah State House as the 1980s approached, noting that the numbers had gone from 8 to three by 1977. She argued that women had a tougher time in politics compared to men. She argued that women with children who had left home should use their time to enter politics: "This is unfortunate [referring to the lack of women in Utah politics]. I encourage women to become politically involved. Women, whose children have grown up are naturals for the Legislature. Yet very few try" (Ferguson 1979). And women had unique challenges to getting the vote out: "'You must be willing to go out and ask people to 'vote for men.' That seems tougher for women to do than men. Women have made political gains throughout the nation in Utah, we've slipped. That disturbs men'" (Ferguson 1979). Women were also not treated with respect by their political colleagues but she did not let those attitudes end her political aspirations: "I can manipulate right along with them. I'm not a member of any natural coalition. I push or pull legislation according to my best dictates. I'm a feminist, but not an 'Uncle Tom' feminist" (Ferguson 1979).

Atwood noted that though she was for women's rights and a feminist, she was not a guaranteed supporter of any directives or political agenda, and mentioned this fact in other publications. She broke with Republicans in the state over abortion laws in 1977 (Hull & Hoffer 2001). Her work was described in an article covering the debate about Utah funding abortions, "Rep Genevieve Atwood, R-Salt Lake, tried to amend the resolution to allow abortions in the case of rape or incest, but failed"(Webb 1977). She also advocated for legislation dealing with equal rights, serving in the house, as she "…has sponsored the Mined Land Reclamation Act and seismic safety legislation, the Open and Public Meetings Act, social services bills

Fig. 8.3 Atwood running. (Ferguson 1979)

dealing with mental health, corrections and alcoholism and drugs, and, in her words, piecemeal equal rights legislation" (Deseret News 1980).

Atwood was also admired by other aspiring female politicians, like Gwen Rowley, who "especially admires" Atwood, even though they were on opposite ends of the political spectrum, with Rogers being a Democrat (Pedersen 1977). Both women were interested in equality for women in society, such as Rowley pictured in Fig. 8.4.

Rowley argued in the article that government needed women like Atwood. She argued that, "'It is essential to get a more equal balance of political power between the sexes in this [Utah] state.' She emphasizes" (Pedersen 1977). Women were working in Utah to support the Equal Rights Amendment, and made it a main political concentration. In Utah women in government were advocating such things as more access to day cares for children, and more quality in employment law (Pedersen 1977). The 1970s saw a type of optimism, even in both political parties, for attempts at equality between men and women (Pedersen 1977).

The 1980s saw a weaponization concerning women advocating for more rights for women and the association with the word feminist, and these examples can be

Fig. 8.4 Gwen Rowley images from article. (Pedersen 1977)

found within the Republican party (Cobble 2003; Spruill 2017; Nickerson 2012; Morris 2022). Attacks on Attwood would later come from within her own party in the next decade, as the Republican party would begin to grow more conservative. Ed Rogers, a fellow Republican, challenged Atwood in the primary of 1980. Atwood's voting district was characterized as "traditionally a moderate to liberal area, and a few strong conservatives like Rogers have won there[,]" according to the political editor for *The Deseret News*. (Webb 1980) Rogers was accused of sending a "misleading letter" to voters in the district (Webb 1980). In the previous election cycle that Rogers won, he characterized then "Atwood-Ferrari" as "extreme liberal feminist" (Webb 1980). Genevieve Atwood was married in 1980 to Ricardo Ferrari, as it was mentioned in *The Deseret News* as the only event that could bring unity to the State House: "The only thing that brought a standing ovation during the budget session was the announcement by Rep. George Richards, R-Salt Lake, that Rep. Genevieve Atwood, R-Salt Lake, was getting married" (Martz 1980).

The political editor criticized Rogers' characterization of Atwood-Ferrari as inaccurate, and that she "….has usually been described as a moderate Republican and fiscal conservative" (Webb 1980). The controversy came when the Senator of Utah at the time, Orrin Hatch, wrote a letter of support for Atwood-Ferrari and Rogers, according to Hatch, became "'very irate, emotional, and upset'" (Webb 1980). Though Rogers claimed that Hatch had apologized about endorsing Atwood-Ferrari, Hatch denied this characterization of events, and continued to maintain his endorsement of Atwood-Ferrari. Rogers wrote a letter that described the Hatch letter as a ruse.

In the letter, Rogers claimed that Hatch was misled into writing an endorsement letter and was bullied by Atwood-Ferrari, whom Rogers described as "liberal feminist" record, which caused Hatch much regret for writing the letter. And when Rogers read Hatch a draft of the letter, the senator did not agree with it, and told him to not send it, which Rogers distributed the letter to the voters. Atwood-Ferrari's husband was interviewed and said that the letter made the difference, as the voters tended to think that the letter was from Hatch. Hatch maintained support for Atwood-Ferrari and thought that she could be the first female governor of Utah. But Hatch turned around and later endorsed Rogers.

Genevieve Atwood, no longer Atwood-Ferrari, would run for the second congressional district representative in 1988. She left politics to head the Utah Geological and Mineral Survey after she did not win re-election to the state house. Atwood, descried as a meteorologist, was praised for her work in the House of Representatives of Utah, but the article referred back to her loss for her bid for the Utah Senate by "a right-wing Republican man," and then she left politics for geology (Dessert News 1988).

Gloria Steinem, who was mentioned in the section about feminism and *Playboy*, had a sister who was a working geologist, and who was vocal about her support for women's rights, though she tried to take a more secondary role than her sister (Pitzulo 2011). In *Lawrence Journal-World* ran a story from the *New York Times News Service* in 1978 which profiled Steinem's sister, Susanne Steinem Patch. The article immediately ran a physical description that was in contrast to the former undercover journalist and celebrity sister, "She drives an orange Volkswagen bus, wears sensibly laced brown oxfords, had all six of her children by natural childbirth – including a young son who has a collection of 565 beer cans – and she has a sister named Gloria Steinem" (Lawrence-Journal World 1978). She was married to a lawyer based in Washington, D. C., and enjoyed disguising her age. The article continued to describe her physical appearance and compare her to her sister: "Across the luncheon table, the voice bore an uncanny resemblance to Gloria Steinem's. With fair hair pulled back into a lose chignon, Mrs. Patch exuded warm, dignity and a low-key sense of humor" (Lawrence-Journal World 1978).

Susanne Patch had a different life track as well, being 52 and was a mother of six children, and was highly active in the PTA, an advocate of natural childbirth, and her work in gemology. Patch has an undergraduate degree in geology and ran a "mail-order gem business." Her geology work included publishing the book *Blue Mystery: The Story of the Hope Diamond*, which was also made into a television special and she was currently attending law school (Lawrence-Journal World 1978; Patch 1976). Patch became interested in geology after collecting rocks in a cigar box over her childhood summers. She got a degree in geology from Smith College, and then pursued employment in the jewelry business, where she presented diamonds across the United States.

In the article she advocated for more rights for women, including the ability to attend law school during the day, as women were not allowed to do so because of the expectation of family and childcare responsibilities. She commented not only on supporting her sister, but Patch expressed her support for the ERA: "The feminist movement for me means full freedom of choice, not a particular course that a person should take…But in order to have true freedom of choice, you have to begin at the beginning—you can't have people say; yes, you can be a nurse, but not a doctor. It was difficult if you grew up at the time when things were different and your options foreclosed" (Lawrence-Journal World 1978). She also seemed to have a supportive partner at home who had encouraged her to pursue her aspirations to become a lawyer. Patch admitted, however, that it was difficult to accept that she had been accepted to law school, as it was a major life change. Now, however, she aspires to pass the Bar.

Her success in pursuing business ventures and law school was through the ability to have additional domestic help, and she cited childcare authorities like Dr. Benjamin Spock, author of *Baby and Childcare*, advocated for mothers of twins to have additional help (Bomback 2022; Spock and Rothenberg 1985; Holt 2014). The idea of "working outside the home on a full time basis when the children were younger[,]" did not occur to Patch because of her desire to stay home and stay with her children (Lawrence-Journal World 1978). One reason for her not working outside of the home is that she wanted to stay out of the workforce and care for the children. But the choice to work outside the home, or work inside the home, was for Patch a personal decision, and a decision for each woman (May 2008; Crouse 2018; Cobble 2003; Bomback 2022). Patch noted that her own decision focused on her own feelings of comfort, "'In some ways, it is easier being at home because you know from having been there with them all last week[, sic] that this is the week they are going to unlock the back door all by themselves" (Lawrence-Journal World 1978). She remained close to her adult children as they still gathered for a dinner on Sunday night. Several women like Patch were making the decisions of pursuing careers, motherhood, and advocating for additional educational opportunities in the 1970s and 1980s, and many were contemplating how they felt about the proposed Equal Rights Amendment (Spruill 2017; Cooble 2003).

Mid Cities Daily News, publishing out of Texas, ran a story discussing how women in geology who were studying in college felt about the Equal Rights Amendment (Mid-Cities Daily News 1975). At Tarrant County Junior College Northwest Campus, Ellen LeMay, was excited to discuss the Equal Rights Amendment. As she perceived its passage would "liberate my soul a little more" (Mid-Cities Daily News 1975). LeMay put the importance of supporting the ERA in historic terms, arguing that women needed to support the ERA. But supporting the ERA was not a rights movement that only supported women but that of "…a people's movement" (Mid-Cities Daily News 1975). LeMay, a woman of multiple identities, such as a "housewife," as well as a student, argued that women were not simply beautiful objects from the Victorian period, "The Victorian age is still here. Some women still feel they should be coddled and have doors opened for them and that's beautiful. It's fine, a nice tradition, but what's so bad about a woman lighting a cigarette, opening a door or getting a job as an executive president of a large major oil company" (Mid-Cities Daily News 1975). These geologists wanted to break free from the Victorian mold of women in science that was discussed earlier in this book with the life of Augusta Kemp.

LeMay acknowledged that not all women were in support of the ERA "...because they think it threatens their traditional roles" (Mid-Cities Daily News 1975). And she compares one of the strongest things a woman could do at the most radical end of supporting the ERA was to be a geologist: "She explained, 'Some women have been suppressed so long that ERA is frightening to them. These women Think Everyone has to go out and be a geologist or everyone has to work for construction'" (Mid-Cities Daily News 1975). She compared the perceived threat of the ERA to a housewife that saw it as potentially turning her whole life upside down.

Some of the male students did support parts of the ERA, such as wage equality but were skeptical that women were discriminated against, or that they were living in a male dominated society. Other men saw the world in 1975 as having sufficient equality, or that society did not truly understand the ERA. One male student, Randy Pruett, was quoted as saying that, "I don't think people and women [blank space] really know what it means" (Mid-Cities Daily News 1975). Pruett stated a dichotomy between women and people. He saw a "men's liberation" angle to it, as men had been taken advantage of with child support, and though the passage of the ERA would be a blow to some men's "ego," passage of the ERA was perceived as an overall good for men by Pruett (Mid-Cities Daily News 1975). Other women interviewed saw positives and negatives about the ERA, but argued against it stripping them of their "femininity" (Mid-Cities Daily News 1975).

The Equal Rights Amendment, and its related discussion, did open up discussions between men and women regarding the role of women in society and if they were restricted by both government, law, culture, motherhood, and society. Women, living within those same political, cultural, and societal changes, brought new questions to their place in the oil industry and geology itself (Klemsrud 1975).

8.4 Social Push Back and Critical Comments About Women in Geology and the Oil Industry

In the January 26, 1978 edition of the *Lewiston Evening Journal*, the paper ran the column of Ann Landers. Landers was an advice columnist for over sixty years and was nationally syndicated to many newspapers. Her column worked where she received a question and then provided answers. Landers received a question from a wife regarding her husband working with a female geologist. The name "Geologist's Wife" is assigned to the wife asking for advice. Geologist's Wife asked,

> Dear Ann Landers: My husband is a geologist for a major oil company. Recently he had to take a young woman geologist out to an oil well to train her. They were together constantly for three weeks, traveled thousands of miles alone in the car, ate all their meals together, even slept out on the rig. (Landers 1978a)

In the late 1970s, women in geology, especially petroleum geology continued to have a robust job growth outlook, but as more women entered into staff geology positions, social and cultural norms were questioned and challenged, but many of the stereotypes and destructive thinking about women and science persisted in society adjusting to more women entering into petroleum geology.

The Geologist's Wife, in a mean swipe, discounted the nature of a potential physical attraction between the two, as she explained she did not believe that the female geologist was a threat: "I'm not worried about the physical attraction because most women geologist are so ugly they could go lion-hunting with a switch. I do resent the proximity between the two of them for that length of time and have hold him so[.]" (Landers 1978a). Geologist's Wife recalled that her husband said nothing unseemly was going on between him and his work colleague.

As the letter went on, Geologist's Wife was also in a mentally taxing place in regard to her attitudes to women's rights movement in the 1970s. She questioned women working outside the home and attacked the female geologist as a proxy for criticizing societal changes asking, "why should women who choose to stay home and be wives and mothers have to put up with such stuff just so these liberated women can prove themselves in a man's field." She also blamed the oil companies for facilitating these uncomfortable working conditions, asking rhetorically "What about the oil companies?" (Landers 1978a). And then answering her question with further criticism of working women: "They profess to care about the welfare of their employees Why not their employees' wives [?] I wonder how other wives feel about these situations and how they deal with their anxieties" (Landers 1978a).

Landers response was to throw cool water on the situation and pointed out to Geologist's Wife that she is very upset, despite her stating that she was not emotional about the situation. The response from Landers was that,

> For a woman who is "not worried" you sound pretty upset. Cool it, dear If your husband's job consists of training new crew members, you'd better accept the fact that some of those crew members are female. What you describe is the result of women's insistence that there be no discrimination on the basis of sex. So, we have to take the bitter with the sweet. (Landers 1978a)

She also responded to the negative comment on the physical appearance of female geologists and tries to distance herself from Geologist's Wife: "P. S. Before I get clobbered by a few thousand lady geologists. I want to go on record as disassociating myself from that comment about their looks. Please don't put the wife's words in my mouth. I make enough gaffes on my own" (Landers 1978a).

A female geologist replied to the column and to Landers. In 1978, the *St. Petersburg Times* ran a letter from "A Woman Geologist in Denver" and the geologist claimed that as she worked as a geologist for four years in petroleum industry, "...I have been treated rather well by most of them male members of my profession." (Landers 1978b) She complained about the perception of her male co-worker's wives. And she hoped that with her writing into Landers it would put issues and conflicts to an end. She pointed out that a lot of co-worker's wives did not realize that they too were not only working and "professional" women, but wives and mothers as well. And though their work makes them happy, their family life does as well. They are not after men or to be better than men, but are there to get the job done: "These women, myself included, are not in the field to 'prove themselves in a man's world' or to seduce their husbands out on a drilling rig" (Landers 1978b). She ended her entry with an attempt at humor and pointing out that oil platforms are difficult places to work, "I suggest that the threatened wives drive out to a drilling rig. They will conclude in short order that it is hardly a place for romance—" (Landers 1978b). Landers added that many of the "scores" of letters she had received from female geologists challenged the previous letter in the same manner, but this letter "…said it best. Bravo!" (Landers 1978b).

In 1977, Dear Abby, another widely distributed national column, received a letter that expressed anxiety about the writer's husband (the writer's initials were given as M. B. from Texas) who worked on an off-shore oil drilling platform that she perceived as surrounded by women. She blamed "Women's Lib" as well as the "government" (Buren 1977). The writer worried that, "There are seven female who are now working side by side with the men on that rig, thanks to the government and Women's Lib!" (Buren 1977). She felt victimized that they all spent so much time together and the government was to blame as it made oil companies hire women. And she stated that her husband cannot do anything to change his circumstances, and her rights as a wife were being violated, "There are plenty of jobs for decent women on land, so why would a decent woman want to work on an oil rig with a bunch of men? They say these women demand equal rights. Where the hell are MY rights?" (Buren 1977). Despite assurances by her husband that there was no extramarital activities and the women were viewed as simply the same as other male employees, she questioned Abby as to whether she believed that was how work was on the oil platform (Buren 1977).

Abby's response was of historical interest because she caste the women working on the oil platform as not engaging with their femininity. This is a departure from the pin-up and glamor culture of the 1940s, 1950s, and 1960s and some early parts of the 1970s that have been discussed in the book. Abby responded that she did side with the husband. And she added an additional response: "And furthermore, any woman who works alongside a man on an oil rig is earning her bread the hard way. If she wanted to cash in on her femininity, I can think of several other jobs she could have chosen"[8] (Buren 1977). Geologists working in the field were increasingly being portrayed as hard working, but an emphasis on their glamor was being downplayed in print culture. Perhaps the emphasis on downplaying the glamor in geology at this point was because of the radical shift in working conditions in the field and the anxiety that other women not working in science felt about the ERA, cultural shifts, or the general women's rights movement.

In 1966, *The Free Lance-Star*, based in oil country in Texas, examined the domestic feelings and lives of women, and the female writer of the article interviewed the wife of a working petroleum geologist. The article "How Unfulfilled Do Women Feel?" written almost a decade prior to the major political movements of the 1970s and 1980s contextualized the deep surface level existential anxieties of women, which likely surfaced during the debates of the Equal Rights Amendment

[8] Here response was originally in Bold font.

era. The article questioned how women feel about their life and how their lives had changed in regards to work. Joy Miller, the reporter, asked a series of rhetorical questions,

> Do American women feel unfulfilled as women, as some critics of contemporary life claim? Has the plethora of mechanical household helpers made time lie heavy on their nondishpan hands? Do they dominate their men at home? If you think it's a yes-yes-yes, the results of an Associated Press survey might surprise you. (Miller 1966)

Jane Culberston, living in Jackson, Mississippi, said that she had no time to feel the anxiety of womanhood. She remarked that, "Good grief, I don't have time to feel trapped or deprived" (Miller 1966). In addition to being identified as "working wife of petroleum geologist and mother of two youngsters" was also listed as "working." Her response to Associated Press (AP) question was labeled as "goodnatured asperity" (Miller 1966). Culbertson did not agree with recent analysis of women's domestic lives as having aspects of "…frustration, entrapment and unfulfillment however over every housewife…" and that she liked her domestic life: "I am most happy to be a housewife" (Miller 1966). She felt both "adjusted" and pleasure in raising children.

But other women did not share her feelings, such as Mrs. Jane Alexander, who was a working lawyer and politician, who also had four children. She was worried at any prospect of working at home in domestic tasks: "Personally, I would feel trapped if I had to stay home and keep house. I don't think a career is every woman's cup of tea, but I enjoy mine" (Miller 1966). Some women remarked that though they did not mind raising children at home, they wished they had more freedom in decisions and other matters. Other women expressed a desire to train for a new career, like teaching, or pursue higher education. While some newly married women felt "disgruntled," and expressed their frustration: "Before you[re married men are very attentive…Afterward, you're take for granted (both by husband and other men)" (Miller 1966).

Women working as housewives also advocated for their own improvement: "No one should be so involved in her home that she lets her mind decay. The mind must remain active…" (Miller 1966) Other housewife wanted to learn about things that they found exciting, or complained about missing out on opportunities of personal improvement because their domestic work kept them too busy for other pursuits. Other women interviewed argued that technological improvements did not make them any less busy. Ruth Cowan made this argument in her pivotal book about gender and household labor. Appliances and household technologies in the postwar period was thought by government leaders and cultural commentators as making household labor easier for women, but it ramped up the expectations, stress, and overall labor as women were not only expected to keep a house pristine and clean, but also produce world class meals, and perfectly laundered clothes (Cowan 1985).

Women noted the inequality in regard to household labor. These complaints were given a voice about the difficulties of household labors: "I think our attitudes about housework have changed now that everyone has to do her own…We no longer expect so much from ourselves. We've learned nothing much happens if we leave the ironing and curl up with a book, or give the floor a quick once-over and get to a committee meeting" (Miller 1966). The article also pointed out the power struggles in marriages in the 1960s. The editorial cartoon included, with both men and women wearing the "pants" or the control of the household, showing the traditional marriage roles of men being in control as in flux, as parodied in the cartoon appearing in Fig. 8.5.

A question was asked toward women as, "Do they feel women have become more dominant in the family relationship with male role accordingly weakened? If married, who wears the pants?" (Fig. 8.4) (Miller 1966). The study found that 90% of women agreed that men had most of the power. Some women's responses

Fig. 8.5 A cartoon that parodied the discussion of power in marriages. (Miller 1966)

included women who did not mind the traditional power dynamic, while others had partners who implied that they could not handle the power of the relationship. But other women were looking for relationships with more equality, not "competition" in their partnerships. These women advocated that no one should have all the power. There were also women who believed that power in relationships was subtler, "Men would rather give in than agree with a woman, and smart women generally use this one weak point in the husband to show their strength. As my husband says 'Give her an inch—she takes a mile,' And I do" (Miller 1966). Other women interviewed complained that they wished they advocated for themselves to a greater degree than for their husbands, while others just admitted that women were in charge. Other women pointed out that sometimes women have hurt their own career aspirations: "I have viewed with alarm the number of demanding young women who have stopped young attorneys' careers in this wonderful new state because 'They didn't like the weather' or something relatively insignificant...[.]" (Miller 1966). The article ended with a note of tragedy, that women were not able to truly relive their talents, as one wife stated from Kansas: "I generally feel women are more dominant and this one of the tragedies of the world [.]" (Miller 1966).

8.5 Stories Critical of the Place of Women in Science and Stories Addressing the Problems of Representation of Women in Geology

During the postwar period not all newspaper stories emphasized glamor and the unlimited advances of geology for women. Some women began to share their experiences of discrimination and struggle. These stories began with discussing the struggles in the field in general, but then started to focus on the struggles with women integrating offshore oil platforms. However, eventually, the glamorous life of the offshore oil women began to appear in magazines, such as *Life*.

In 1959, the *Toledo Blade* ran a story profiling "Mrs. John W. Kitchen" who was a working geologist. The article opened with the narrative that Kitchen was defying expectations for women: "On VIRGINIA STREET, there's a little woman whose ordinary working day might be a bit discouraging to women in other professions" (Kerstetter 1959). Kitchen was associated with a "binocular microscope" and working in the field to search for "its chemical riches" (Kerstetter 1959). The article focused on her appearance: "Unlike the working woman shoe work – a – day apparel might include a prim suit and sports coat, Mrs. Kitchen wears high-laced field boots, wool blouses and jackets and blue jeans" (Kerstetter 1959). This was keeping with the pin-up image of the woman, working in men's clothes in the field that is just another employee (Hillman 2015). Her transportation was a jeep, and she often had to walk through "ankle-deep mire" in Michigan (Kerstetter 1959). At the time of the article, she was working as a senior geologist for Sun Oil Company. Her job mostly focused on constructing maps where oil deposits might be accessed in the Michigan area.

Kitchen was also highly educated, like the other geologist profiled in the post-war. She was using her geologic knowledge to determine which changes in the earth indicated oil deposits,

> To do this, however, one must be rich with information about the earth and its life as recorded in rocks, such as igneous rock, which according to Mrs. Kitchen, is formed by solidification of a molten magma, or the metamorphic layer, which is pronounced change generally affected by pressure, heat and water, resulting in a more highly crystalline condition. (Kerstetter 1959)

The article praised how exciting geology was for the reader. But the article changed where instead of advocating the career for women, it shifts to deter them from entering the field. And the industry as a whole, according to both the interview subject and the article, report that there is not a lot of opportunity for women. She appeared with geological "specimens" in Fig. 8.6. Instead of praising her exceptional nature in the field as a way to build excitement and recruit women into the profession and oil industry, it is seen as a negative. The article read as a cautionary note to the reader, "However, the prospects of a woman succeeding as a geologist

Fig. 8.6 Kitchen posing with geologic samples. (Kerstetter 1959)

today are dim, and Mrs. Kitchen, who is the only woman geologist in the petroleum industry in Toledo, has some discouraging news and a bit of terse advice: 'Girls, chose an easier profession'" (Kerstetter 1959).

Women needed lots of education to be competitive, as the field was becoming highly technical, and she recommended that those students aspiring to work in geology to plan for a college education that was not simply four years, but eight years long. Oil companies wanted new employees to study for a doctorate. Prior to Kitchen's employment, entry into the profession and oil industry really only required a bachelor, and geologists had more bargaining power in gaining employment. But now there were around 2000 graduates in geology looking for jobs, according to Kitchen, and there were only 700 open spots. But she did emphasize that the work in water-based geology (hydrology) led to the improvement of the natural world, as water was getting "scarce" because of air conditioning technology, and people were starting to pay attention.

Kitchen, who is never identified by her first name, gives a narrative that is against the advertising and propaganda by oil companies, the statistics, and the testimony of other geologists around the same time. However, Kitchen represents one of the first critical testimonies from female geologists against the oil industry, beyond the occasional complain about sexism or discrimination in the field, such as the write up that included the New York geologist from Flatbush working in Texas Oil. Critical reports of the oil industry would begin to emerge alongside the increase of the women's rights movement in the late 1960s and early 1970s. But the oil industry would continue to promote women in the profession and opportunities for women. Many articles included tacit examples of women "fighting" for a place in the field.

8.6 Conclusions

There were new questions in the general field of science regarding women's participation. Much of this debate was heated up with the Equal Rights Amendment. The 1979 edition of *The New Scientist* discussed the historic nature of the American Association for the Advancement of Science meeting in Houston, as it had been moved from its planned city of Chicago (New Scientist 1979). It was moved because the state of Illinois had not supported the passage of the ERA. This was also the first meeting that had a planned session about women in the field of science. The session was summarized with the idea that "Science is not value free[,]" likely stated by one of the presenters, and the summary included the ideas that men had power in the world, and this framework included scientific practice and that male domination of science was "now under challenge" (New Scientist 1979). Women challenging the male domination of science could be as important as the Copernican Revolution in astronomy. Some sociologists-scientists argued that human pre-history had a more of a community, or equal distribution of power, as demonstrated by savanna baboons, and that it might be good for the world in general to return to this more equitable

Fig. 8.7 Cartoon from the AAAS publication satirizing women in science (New Scientist 1979)

structure. And adding more "concerns" of the women's movement might improve science's social and intellectual value system (New Scientist 1979).

A cartoon, Fig. 8.7, was included with the summary of the AAAS panel regarding women in science.

The cartoon seems to imply out of touch male scientists attending the meeting, only thinking that there could be male "masters" of science. Other interpretations of the cartoon could include a more sexualized component of science, as the word "mistress" implied an illicit extra-marital sexual relationship. The title of the summary of the AAAS meeting was also loaded with meaning, "Feminists' new world apes baboons" (New Scientist 1979). The association of a new world with apes likely did not have a positive connotation, given the history of associating people with apes (Haraway 1990). Also the inclusion of the term "apes" implied the verb meaning the copy in a less than impressive manner. Ideas about gender are both mapped on and explored metaphorically through primatology, as many gender scholars have pointed out that ideas about sexism are applied to interpreting the behavior of primates. During the 1980s, historians began to criticize the sexism inherent in primatology (Haraway 1990).

While the American Association for the Advancement of Science was coming to terms with, or perhaps bristling against, women's greater participation in science, female geologists continued to reflect about the role of women in science. In a profile covering Dr. Dianne R. Neilson, a geologist who worked as a consultant geologist and owned her own business in Salt Lake City prior to turning thirty years old reflected on the place of women in science.

Neilson, pictured in Fig. 8.8, was interviewed and was quoted that, "....women in geology still are not numerous but they don't encounter much prejudice anymore from oil and mining companies" (Pope 1979). Though women, especially those I

Fig. 8.8 Dr. Dianne R. Neilson. (Pope 1979)

will discuss in the next chapter, commented more on discrimination, sexism, and racism in coporate culture. These women commented that oil companies had problematic racist corporate culture, and their thoughts on these matters appeared in newspaper and magizine profiles, while still containing aspects of the older glamour articles of the attractive working female geologist. These profiles, though containing more social commentary, continued to emphasize the older ideas about the glamour covered in the postwar period. The transitions were not clean and did not occur rapidly, and women differed in the interpretation of their experiences in petroleum geology. Neilson added that, "'Nor have I ever had any difficulty working with men in the field.'…'in spite of all our modern instrumentation, geology still is field work. You still see a few prospectors with a burro and a few tools. They are amateur geologists but they have made some of the great mineral finds. The professional geologist does his or her most important work out in the hills, chipping away at rocks and studying them, and inspecting the cores the drillers bring up'" (Pope 1979).

The article commented on Neilson's appearance and clothing. Her profile was much like that of a glamorous World War II geologist. The article described her work in the field and the necessary clothing: "That means wearing blue jeans, rough boots and a hard hat and driving your own jeep on the most rugged mountain trails., It also means a lot of hiking alone with a black on mountain slopes and desert plains" (Pope 1979). All the field hardened drills respected Neilson, as the article reported that, "The drill crews are pretty tough babies as a rule, but Dianne says she's never had any difficulty with a driller. They all treat her with respect and accept her as a competent scientist" (Pope 1979). Her physical features were also described in the article as "a rather striking redhead." But a diversion from the older narratives explored in this book is that Neilson is supported by her husband and free from domestic labor: "'My marriage made it a lot easier to establish my practice,'… 'I didn't have to worry about the rent or the phone bill.'" It was her career that was in the driver's seat and her husband supported her financially with a stable job that allowed her to pursue international geological projects, such as in Italy. She also was not obligated to have a family, and brushed off the usual narrative of a woman having children, as the article reported that, "Asked about the children she just smiled and said, 'Maybe later'" (Pope 1979).

She was also openly ambitious. She wanted to move away from work with nuclear energy and uranium and work in the oil industry. When she was asked by the reporter as to what was motivating her to make the jump into the oil industry and petroleum geology, as well as valuable mineral exploration, she responded that "Because that's where the big money is—and the nation's future." (Pope 1979).

Though profiles throughout the 1970s and 1980s like Neilson's contained many of the ultra-feminized and glamorous geologist at home in the oil fields, other profiles of female geologist began to offer another perception and commentary on the practice of petroleum geology and offered more critical reflections of the experiences of women in science. Interestingly, both profiles would further explicate the experiences of women in geology. Changing ideas about the roles of women in society, relationships, and science put forth new ideas and perception of glamorous women in science, and provided both positive and negative aspects of their stories.

References

(1950) Schedule woman geologist for broadcast Tuesday. Prescott Evening Courier 15 June
(1975) Equal rights amendment: voices from the college scene. Mid-Cities Daily News 7 March
(1978) Steinem sister stays out of shadow. Lawrence-Journal World 23 February
(1979). Feminists' new world. New Scientist 81
(1980) Atwood-Ferrari seeks Farley seat. Deseret News. 10 May
(1988) State geologist considers gop run against owens. Deseret News 26 January
Bomback A (2022) Long days, short years: a cultural history of modern parenting. The MIT Press, Cambridge
Bronstein C, Strub W (eds) (2016) Porno chic and the sex wars: american sexual representation in the 1970s. University of Massachusetts Press, Amherst
Van Buren A (1977) Dear Abby: She wants women to stay ashore. The Sumter Daily Item (5 March)
Clerici PC (2014) Boston marathon history by the mile. The History Press, Charleston
Cobble DS (2003) The other women's movement: workplace justice and social rights in modern america. Princeton University Press, Princeton
Cooke J (2021) Come fly the world: the jet-age story of the women of pan am. Houghton Mifflin Harcourt, New York
Cowan R (1985) More work for mother: the ironies of household technology from the open hearth to the microwave. Basic Books, New York
Crouse ER (2018) America's failing economy and the rise of ronald regan. Pelgrave Macmillan, New York
Dines G, Jensen R, Russo A (eds) (1998) Pornography: the production and consumption of inequality. Routledge, New York
Ferguson G (1979) Running is way of life for Rep. Atwood. Deseret News (1 February)
Fraterrigo E (2009) Playboy and the making of the good life in modern america. Oxford University Press, New York
Gotass T (2012) Running: a global history. Reakiton Books, London

Graff GM (2022) Watergate: a new history. Simon & Schuster, New York
Haraway D (1990) Primate visions: gender, race, and nature in the world of modern science. Routledge, New York
Harrington C (2020) What is "toxic masculinity" and why does it matter? Men Masculinities 24(2):345–352
Hillman BL (2015) Dressing for the culture wars: style and the politics of self-perception in the 1960s and 1970s. University of Nebraska Press, Lincoln
Holt MI (2014) Cold war kids: politics and childhood in postwar america, 1945–1960. University Press of Kansas, Lawrence
Hull NEH, Hoffer CP (2001) Roe v. wade: the abortion rights controversy in american history. University Press of Kansas, Lawrence
Huessy F (1981) Playmate riding a wave. Spokane Daily Chronicle (11 February)
Katz JN (2007) The invention of heterosexuality. University of Chicago Press, Chicago. Reprint of 1995 edition
Kerstetter E (1959) A woman in geology: it's a fascinating profession, but it isn't recommended. Toledo Blade. 15 February
Klemsrud J (1975) Photo-journalist likes to interpret women who work. Ocala Star-Banner (14 August)
Kutulas J (2017) After aquarius dawned: how the revolutions of the sixties became the popular culture of the seventies. University of North Carolina Press, Chapel Hill
Landers A (1978a) Ann Landers: geologist's wife. Lewiston Evening Journal (26 January)
Landers A (1978b) Ann Landers: female geologists. St Petersburg Times (25 March)
Levine E (2007) Wallowing in sex: the new sexual culture of 1970s american television. Duke University Press, Durham
Martz M (1980) The house really hit the roof! Deseret News. 12 February
Matelski EM (2017) Reducing bodies: mass culture and the female figure in postwar america. Routledge, New York
Maxwell A, Shields T (eds) (2018) The legacy of second-wave feminism in american politics. Pelgrave Macmillan, New York
May ET (2008) Homeward bound: american families in the cold war era. Basic Books, New York
Melnick RS (2018) The transformation of title ix: regulating gender equality in education. Brookings Institution Press, Washington, D.C
Miller J (1966) How unfilled do women feel? The Free Lance-Star. 11 August
Morris R (2022) Goldwater girls to regan women: gender, Georgia, and the growth of the new right. University of Georgia Press, Athens
Nickerson M (2012) Mothers of conservativism: women and the postwar right. Princeton University Press, Princeton
Osgerby B (2001) Playboys in paradise: masculinity, youth and leisure-style in modern america. Berg Publishers, Oxford
Patch SS (1976) Blue mystery: the story of the hope diamond. Smithsonian Institution Press, Washington D. C. Also see the entry in the Smithsonian archive about the book and the exhibit: https://siris-sihistory.si.edu/ipac20/ipac.jsp?&profile=all&source=~!sichronology&uri=full=3100001~!3687~!0#focus
Pedersen RM (1975) "Women power" has become mighty force in politics. Deseret News. 4 January
Pedersen RM (1977) Mention politics and Gwen lights up. Deseret News. 14 April
Pitzulo C (2008) The battle for every man's bed: playboy and the fiery feminists. J Hist Sex 17(2):259–289
Pitzulo C (2011) Bachelors and bunnies: the sexual politics of playboy. University of Chicago Press, Chicago
Pope L (1979) Field trips were fun—now they're her job. The Deseret News (18 May)
Reichert T (2003) The erotic history of advertising. Prometheus Books, Amherst, New York

Spruill M J (2017) Divided we stand: the battle over women's rights and family values that polarized america's politics

Sallquist B (1980) A beauty bares views on nudity. Spokane Daily Chronicle (11 April)

Sears ES (2015) Running through the ages, 2nd edn. McFarland & Company, Inc, Jefferson, N. C

Spock B, Rothenberg MB (1985) Baby and child care. 40th Anniversary Edition. Pocket Books, New York

Switzer K (2017) Marathon woman: running the race to revolutionize women's sports. Hachette Book Group, New York

Tumin R (2022) In 1972, only 8 women ran the race. Today 12,100 are running New York Times. 18 April. https://www.nytimes.com/2022/04/18/sports/women-running-boston-marathon.html#:~:text=ran%20the%20race.-,Today%2C%2012%2C100%20are%20running.,a%20change%2050%20years%20ago.&text=For%2076%20years%2C%20men%20were,running%20of%20the%20Boston%20Marathon

Vantoch V (2013) The jet sex: airline stewardesses and the making of an american icon. University of Pennsylvania Press, Philadelphia

Webb LV (1977) Panel votes to ban funds for abortions. Deseret News (9 February)

Webb L (1980) Case of misleading letter: whose fault? Deseret News. 16 September

Wdowik M (2017) The long, strange history of dieting fads. The Conversation 6 November

West M (2020) The birth of the pin-up girl: an American social phenomena, 1940–1946. Ph.D. Dissertation, University of Iowa

Wohl R (2015) Up in the air: new approaches to the history of american aviation. Rev Am Hist 43(4):687–696

Chapter 9
Resistance, Discrimination, and the New Glamour

9.1 Introduction: The New Glamour

Glamour profiles of women working in geology included more accounts of their struggles and discrimination because of their gender. Many of the profiles could be considered "Bucking the System" and fighting against the status quo of promoting female geologist because of their physical and ultra-feminine features. Female geologists lived in the US culture and were affected by changes in ideas about marriage and what it was to be a woman in society. New books about fashion and beauty reflected new ideas about freedom for women. New books about fashion emphasized wearing things like pants to the workplace as part of being the new woman. Books about beauty also included ideas about how to recover energy, as women were working hard in offices and their jobs required them to rest.

Newspapers not only covered the new working geologist who was not afraid to call out discrimination, publications included geologic work performed in the African-American community. Black focused magazines like *Ebony* began to profile working female geologists in the same manner as white women in the World War II era (Brown 2020). Other profiles about working geologists expressed the struggles of working in the oil industry and how destructive the market could be, such as in the 1980s. Women in the 1960s onward were also looking to enter management positions in geology. Glamourous profiles of female geologists leaned into accounts of discrimination and expressed how difficult it was to be a woman in geology. Many profiles included the accounts of discrimination and struggle, and accounts of physical beauty were replaced with these personal histories: which redefined the new glamour of geology. The new glamour consisted of accounts of transcendence of women working in geology despite the difficulties of sexism. Profiles from the oil industry and popular magazines continued to emphasize the adventurous and glamourous nature of being a working female geologist. The new glamour emphasized the changing nature of female identity.

E. A. Driggers, *Glamour and Geology*,
https://doi.org/10.1007/978-3-031-64525-9_9

9.2 Resisting the System and Female Introspection: Making Their Own Rules

The resistance narrative has been a narrative that has been one of emphasis by historians. First and exceptional women are themes that emerge in the studies of Robbie Gries and Elizabeth M. Bass (Gries 2017; Bass 2020). But what about women who are somewhat in the middle. They love and embrace their jobs, but often do not know how they feel about women's rights, are not committed to complete revolution in the field, or they are benefited by fitting into the glamour-based geology and oil system. Sometimes it makes sense to fight for rights, while other women believed they should bide their time as conditions might change. The idea of resisting the system was of benefit for women geologists today as they continue to open the field to women, but as Bass would later move away from that narrative in the end of her introduction of her dissertation, what about the rank-and-file women (Bass 2020). I would argue that bucking the system became the common theme of the oil and gas industry, and this will continue into today, as geologists, as well as women in the United States, created identities that were independent and individualistic.

An example of women seeing themselves as revolutionaries for women's rights can also be seen in the coal and mining field, for comparison. Sandra Thielen was profiled in *The Southeast Missourian* in 1971. Thielen was known as "Sam" as that name was included on her hard hat. She embraced her revolutionary place working in the coal mines, as the article ran the title "Sam bucks bastion in mining field" (The Southeast Missourian 1971). Her physical features were described as "…a 5-foot-4½, 115-pound senior at Colorado School of Mines, who wears her light brown hair long and straight" (The Southeast Missourian 1971). This description was much in keeping with the pin-up profiles of the 1940s and 1950s. When Theilen graduates, she will be the third female graduate in the school's close to one-hundred-year history. The article embraced her spirit of excitement for breaking through a glass ceiling: "She's serious about a career in mining even though the profession has remained a closely guarded bastion of males" (The Southeast Missourian 1971). Laws had to be repealed in Colorado for Thielen to be able to work, "Superstition is rife about bad luck supposedly befalling the mines into which women venture. In fact, it was only this year that the Colorado Legislature repealed a 19th century law prohibiting women from working in mines" (The Southeast Missourian 1971).

The theme of bucking the system continued in the article as the author rhetorically questioned why she was interested in challenging the system. Thielen's response was recorded as a strong volunteering to fight the system: "'Someone has to open up the field,' she says" (The Southeast Missourian 1971). Her field work training included working with thirty-five men in the "school's experimental mine at Idaho Springs" (The Southeast Missourian 1971). She noted that the work was "dirty" but she, like the testimony found in the advice columns mentioned above had the experience of being treated somewhat equally: "Guys consider me just one of the guys. They're just great about it" (The Southeast Missourian 1971). But the narrative then included the fact that Thielen had formed a romantic relationship with one of the men in her program, but this had occurred prior to the writing of the article, and he was credited with spurring some of her interest in geology, though she had been interested in majoring geology prior to meeting her boyfriend. Her experience with him, and his encouragement to take an introductory course, caused her to switch majors to emphasize mining. The article, featuring a cartoon that appears in Fig. 9.1, conveys the ideas of changing identifies for women in culture and science. The article then ended on an optimistic note, as the article mentioned that a couple of women were joining the major each year, "The way things look, Miss Thielen won't be a lonely woman miner for long. Each of the four undergraduate classes at Mines now has a woman mining major, she said, and the total enrollment of women is rising rapidly at school, which has always been coeducation" (The Southeast Missourian 1971).

The article was then followed by another article run from New York that explored the duality and pressures of beauty standards and fashion for women, entitled "Beautiful woman recognizes self" (The Southeast Missourian 1971). The article continued to explore the tensions in women dressing for work, as the previous conventional trend had been in conformity to dominate fashion trends in order to be considered "beautiful" (The Southeast Missourian 1971). And there was considerable stress for women "bucking" trends in fashion: "Women not suited for hotpants longed for them wistfully. Young girls with curly hair pulled, tugged and ironed their tresses to get them straight, only to find that their now straight hair was out; curls and waves were in. Skirt lengths have been jerked up and down so many times that men's heads are spinning" (The Southeast Missourian 1971). Women were spending large sums of effort, time, and money to keep up with fashion.

The article argued that women then were not in control the fashion "...of the moment was wearing her" (The Southeast Missourian 1971). Each woman was stuck hoping that the next cycle of fashion would be something that "suited her[.]" The working woman, the article stated, did not have to wear clothes and hair styles that were simply deemed fashionable, and should pursue a look that she preferred for her own comforts: "She did not belong with her new clothes or her new hairdo.

Fig. 9.1 Cartoon that appeared above the story about Theilen and in the story "Beautiful woman recognizes self". (The Southeast Missourian 1971)

If she did not give up on her beautification program after this frustrating attempt, she waited hopefully for the next style, which may have been equally unsuitable for her" (The Southeast Missourian 1971). In a way of revolution, the article argued that there was a way out of the cycle in 1971.

All fashion, as well as beauty, should begin with the woman, "It starts with the individual woman, herself—the place perhaps, where it always should have started" (The Southeast Missourian 1971). How a woman styled herself represented many facts of her identity in the 1970s: "A woman's clothing and grooming present to the world an image of the kind of person she is" (The Southeast Missourian 1971). In order to properly style herself she needed to know who she was in the world, and then judge, based on her attributes, what was the proper style "In order to project her true identity, she must know who she is. Then she needs to honestly evaluate her face, her figure and her lifestyle before deciding on her 'look'" (The Southeast Missourian 1971). Styling and beauty were part of life, not external from life, and needed to be integrated into the daily workplace. The article gives the example of the activities of the day, "If Sally is an executive secretary, back, slinky dresses will do only after five. If Susie is a schoolteacher, hotpants won't do, but skirts, sweaters and basic dresses will" (The Southeast Missourian 1971).

The article advertised its approach to fashion as revolutionary and liberating. The independence that picking one's clothes, regardless of fashion styles, made women more independent, "This new approach to fashion and beauty lets a woman wear what she likes, not what someone else has told her she must like" (The Southeast Missourian 1971). However, standards are still clearly defined with the prescriptions mentioned above. Office work and clothing became a part of the conversation about beauty, womanhood, professionalization, and science. Women were also voicing their ideas, preferences, and independence at the office, as this history was discussed in chapter seven. These trends continued and books about fashion discussed self-care and the challenges of office work.

The article referenced the book, written by Donna Lawson and Jean Conlon, entitled *Beauty Is No Big Deal*. The book, summarized in the article, wrote that it was a "common sense" manual essentially to styling the independent woman. Mirrors were evoked to plot the best outfit, "The authors suggest either standing in front of a mirror to see oneself at all angles, or getting a photographer to snap photos from all angles: Is one's posture good?" (Lawson and Conon 1971). Though the book advertised a revolutionary approach to fashion, it still exposed its own beauty standards and advocated the idea that there was a perfect outfit for a woman. Women needed to look at the mirror in order to judge their whole body for its suitability, "Is one's hairstyle too flat? How are the legs? What is the best feature? What is the worst? Sugar-coating the appraisal defeats its purpose. The purpose is to see what one has to work with and what needs to be worked on" (Lawson and Conon 1971, pg. 118).

The book, published in 1971, featured many drawings and illustrations, and some of them appear in Figs. 9.2 and 9.3. The book had sections emphasizing that women's fashion should be comfortable and women working in the office should have the ability to have a break from their work or fashion shoes. But the book also

Fig. 9.2 Illustration from *Beauty Is No Big Deal.* (Lawson and Conon 1971, pg. 124)

emphasized diet culture and fatphobia of the 1970s (Lawson and Conon 1971; Matelski 2017; Wdowik 2017). The book would advocate relaxation with an emphasis on how to "slim down ankles," pointing out to working woman that they needed to "Give your feet a break. Change shoes at least once a day. Keep an extra pair of shoes on hand at the office to wear when you go out after work, change into sandals or slippers when you get home. Better yet, lie on your bed or the floor and put your feet up against the wall. Feel the fatigue of the day drain from them. Groovy" (Lawson and Conon 1971, pg. 118). But the article pointed out that both writers hated exercise and wanted to provide a space for women to complete exercise without having to take extra time out of their day, or while sitting, to pursue an exercise regime.

Aerobics and exercise became incredibly popular during the 1970s and 1980s, increasing the pressure and demands on women's physical appearance (Friedman 2022; Black 2013). Though women were being opened up to another set of physical beauty standards, the authors argued that the book was empowering because the book allowed women to be individualistic, though it was advocating prescriptive physical standards. But the book also advocated an idea of the importance of internal beauty: "Further, what makes a woman beautiful is her inside as well as her outside" (Lawson and Conon 1971).

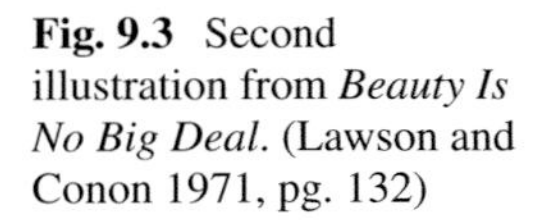

Fig. 9.3 Second illustration from *Beauty Is No Big Deal*. (Lawson and Conon 1971, pg. 132)

The women advocated as well that environment has a lot of effect on the mood and whether a woman was "beautiful," and they argued that if a woman was having a bad day, changing around their physical environment would benefit the worker:

> Donna and Jean [the authors of the book] point out that environmental has a lot to do with how a woman feels and, therefore, with how she looks. A woman having an off – day rarely looks beautiful. They suggest decorating home and office to one's own needs—whether it means orange walls and wild daisies strewn about or pale pastels subtly blended. (Lawson and Conon 1971)

And the women wanted balance, and the article tried to consolidate the revolutionary ideas of how women could imagine themselves and their fashions: "Fashions and cosmetics should serve women, not rule them. All women need to do is make their own rules for the beauty and fashion game" (Lawson and Conon 1971).

9.3 Cultivating Diversity and Celebrating Women in the Petroleum Industry and Positive Messaging

Newspapers in the African-American community reported on African-American women getting professional jobs and participating in geology. The 1963 edition of *The Afro American*, a newspaper that reported to the African-American community, ran an article that promoted the fact that "Professional Workers include more women" (The Afro American 1963). Though the numbers had increased, however, they were in the same proportions as they were in 1940. However, there were some increases in women working in geology, "Small numbers of women were also engaged in the professions considered relatively unusual for women, such as engineering, architecture, geology, the ministry, and forestry" (The Afro American 1963). Women increased in the scientific and engineering fields and this point was highlighted in the article: "OF PARTICULAR interest [?] view of our present day emphasis on scientific and engineering skills, were the relatively large increases in women scientists, technicians, and engineers – evidence that women are sharing in the expanding demand for qualified personnel" (The Afro American 1963).

Working women geologists were also active in the African-American community, such as Katherine Greacen Nelson (Kluessendorf and Mikulic 2018; Paull and Paull 1984). Nelson made periodic presentations at the Baltimore North Side YMCA (The Afro American 1954). Nelson worked with African-Americans at the YMCA as the leader of teenage groups, after she concluded her professorship in geology at Milwaukee-Downer College. Nelson was identified as the first women to earn a doctorate in geology at Rutgers (The Afro American 1954). In chapter four, I examined the problems of geological education caused by a lack of resources and faculty who could teach in this area in historically black colleges and Universities, according to scholars of the 1940s (Schneiderman 1994). The lack of resources and geology faculty caused there to be a small amount of geology majors produced at historically black colleges and universities (HBCUs) in the time period of segregated education. The underfunding of geology education caused a cascade of a lack of African-American participation in geology. It was likely that there was little access for interested African-American students to enter into Ph.D. programs, which would have facilitated increased geology teaching and undergraduate programs at HBCUs because of segregation at most US schools (Slater 1994). African-American women experienced continued hardships in participating in geology and working in geology because of racism, discrimination, and sexism, and black women's participation in geology is still one of the most limited groups (Farinde and Lewis 2012; McCandless 1992; Hanson 2009). Community engagement and outreach in black communities seemed to help the situation. Nelson was an advocate for geology across many communities, including African-American communities. Her classes in geology were advertised in other issues of black newspapers, like *The Afro American* in Milwaukie. Her geology course was advertised along with bridge classes and hat-making classes (Travis 1956).

African-Americans experienced discrimination when they did work in geology. An example comes from New York. Ruby Bell was working as a geologist in the Office of the United States Weather Bureau and she filed a "complaint with the Fair Employment Practices Committee (FEPC)" in 1943 (Bloch 1965). The complaint stated that she was denied a job because she was African-American. Bloch, a historian writing about the discrimination case described Bell as, "…a Hunter College graduate with graduate work in geology and meteorology, was referred to the Weather Bureau by the United States Civil Service Commission" (Block 1965, pg. 363). She was interviewed by the Weather Bureau and was awaiting notification of employment and training information. However, she performed some "private investigating" and found that there were no African-American employees at the Weather Bureau, and that white applicants, found on the same list that her name had been included on were hired before her, and were less academically qualified. Some employees had no training at all. The FEPC acted and told the Weather Bureau to hire Bell as a "junior observer." But the Weather Bureau did not follow the direction of the FEPC. And a joint review of Bell's case found that she was discriminated against for the position involving a Mr. Swain who was in charge of personnel at the US Department of Commerce, "A joint conference was arranged between FEPC's officials and G. Swain, chief of the Division of Personnel of the U. S. Department of Commerce. After Mr. Swain examined the evidence, he agreed that this was an instance of discrimination because of color[5]" (Bloch 1965, pg. 363). Though the outside review found that she was discriminated against, Bell did not ever receive employment.

By 1997, newspapers were running Associated Press stories about the large amount of lawsuits that oil companies were paying for discrimination cases. In the *Kingman Daily Miner*, the story "Oil industry is reeling from sex and race cases" was published on January 6, which described the one hundred and seventy-five million dollars Texaco had paid out to race-based cases of discrimination (Kingman Daily Miner 1997). There was another 100 million dollars Royal Dutch Shell might have to pay out if they lose or settle the additional cases that were yet to be adjudicated. The image of oil companies was under threat in the 1990s. The article contextualized the problems the companies were facing, "The problems at Texaco and Shell, along with Chevron—which has paid millions in recent years to settle discrimination complaints—are not isolated cases in the oil industry, according to the head of a consulting firm that tracks minority performance in business." The industry was experiencing wide-scale problems in regard to minority employment. Compared to other industries, the oil industry was often at the bottom for minority and female employees in regard to "…senior management, recruitment and entry-level positions…" (Kingman Daily Miner 1997). Spokesmen for the oil industry stated that they do not actively discourage minority and female employees, but the industry itself has been slow to create diversity programs or change its image.

The industry itself had problems in its internal culture, as the article described two examples of racism and poor company culture, "Texaco agreed to the record settlement payment after a disgruntled former executive released a tape in which top company officials were heard using racial slurs, mocking the black cultural festival Kwanza and plotting to hide or shred documents sought by the plaintiffs." Another example was cited from Royal Dutch Shell. The article noted that "…a group of black workers is suing for $100 million, alleging a system of racial and sex discrimination. A second group of Shell employees filed suit last February charging racial discrimination by the company" (Kingman Daily Miner 1997). Some of the named workers, such as Wilson Jackson, argued that they were systematically discriminated against, and Wilson criticized the industry saying, "…the company has held down black employees by withholding deserved promotions and merit awards and has lowered positive evaluations for black workers. They also charge Shell failed to reprimand workers for racist conduct at work." Wilson continued with criticism of the industry and its racist culture, "Caucasians…move up faster than black who do the same or a better job[.]" (Kingman Daily Miner 1997). Stories about racism and sexism issues in the oil industry continue to appear in the news even to this day.[1]

Though there was a culture of sexism and racism in the oil industry, black women did gain employment and were profiled in African-American focused magazines. In 1984, *Ebony Magazine*, a US-based magazine that covers African-American life-style, culture, and current events, ran a story featuring Edith Williams, in its article "Lady Pioneer in the Oil Fields" (White III 1984).

The article, featuring many photographs of Williams working in the field, appears in Figs. 9.4, 9.5 and 9.6. It promoted William's work with Mobil in searching for oil wells. She is pictured gathering information in Wyoming from geophones that show geologists information about waves that are affecting the land. The article opens with the recording of William's perfume being spread through the winds of the plains; the article makes note that Williams is also "...one of the first Black American women to earn a degree in geophysics..." and was working to help Mobil Oil find oil around those same plains (White III 1984).

Noting her fear of heights, she continued through her work, scaling mountain sides to gather instrumentation. Measuring these vibrations are important in the search for oil, as it shows how viable well sites are for oil and gas wells, as driller wants to avoid explosions and difficult well sites. At age 25 at the time of the article, she was working in the oil field and had already completed a Master of Science in geophysics at Stanford. The article also commented on the exceptionalness of black geologists working in the oil fields in the 1980s, "Black geophysicists are, of course,

Fig. 9.4 Williams working in geology. (White III 1984)

[1] For example, see the 2017 story that appeared on NPR's "All Things Considered:" https://www.npr.org/2017/11/05/553969144/big-oil-has-a-diversity-problem

Fig. 9.5 Williams working in the field. (White III 1984)

more rare; possibly they can be counted on a pair of hands. Aside from the challenge of breaking into a field being one of the first Black women to earn such a degree does not occupy much of her thinking" (White III 1984, pg. 60–61).

Williams was quoted in response to the article's point of her being one of the few black geophysicists working in the oil fields: "'I've been in that situation (first black or first woman) basically all my life,' she says. 'I've just been blessed since the day I was born. I'm not saying that the road to where I am today was so easy; I've had my trials and tribulations just like anyone else. But the fact is I've met so many human beings who didn't have to take an interest in men, but did'" (White III 1984, pg. 62).

Williams went on to comment on racism and the academy. She noted that, "'I was very comfortable in the academic world,' she adds. 'Racism isn't as prevalent

Mobil's 12-month Explorationist Development Program includes 11 weeks of applied geoscience, which culminates in a presentation by Ms. Williams at her Dallas base (right). Below, she confers with Mobil staff geophysicist Reginald Beasley.

Fig. 9.6 Williams giving a presentation at Mobil. (White III 1984)

in the academic community as it is in the real world. I'm learning to live in the real world as an anomaly. Sometimes I think everybody's a mono-color; we're all one. And I just think of myself as another human being'" (White III 1984, pg. 62). While in Wyoming, Williams was completing a year young program with Mobil that eventually led to her becoming a "supervising geophysicist." The article analogized the program to a medical residency. She was in the field collecting data at the time of the article, and then she planned on returning to Dallas for around four months to analyze the data.

Williams was inspired by Jacque Cousteau, a famous twentieth-century ocean documentarian and explorer, and was planning to pursue the geology of the ocean, but found herself in the oilfield after a summer job at the US Geological Survey. Part of her job that summer required her to travel to Cape Cod, Massachusetts, where she performed geophysical work. She became further interested in geophysics when she learned, "...two Blacks, Roland Henderson and Randolph Bromery, were working in that field" (White III 1984, pg. 62). Williams was able to talk to Henderson and he promoted geophysics as "...the future..." (White III 1984, pg. 62).

Williams had her eye to the future and towards space. Reflecting on what Henderson had told her, she thought that, "I took it a step further. We're running out of natural resources. I was thinking of extraterrestrial mining. A colony mines the minerals, puts them on a shuttle and brings them back to earth. I plan to be there, an onboard scientist conducting experiments" (White III 1984, pg. 62). She excelled in her geology studies at the University of Miami. She received a high salary offer from a company specializing in oil production, and they wanted her to study geophysics. Williams then pursued a masters in geophysics at Stanford. Williams also praised her parents and their support. Her father was a hardworking employee at a factory, and her mother was first a nurse, then transitioned into education. They made sacrifices, such as putting all of their three children through private school and emphasized the value of hard work. She remembered the lessons that they taught her: "'They taught me to have self-esteem and to work hard, but the most important thing they stilled in me was God,' Ms. Williams says. 'I'm traveling through territories travelled by none others like myself; my company is betting millions on the integrity of my expertise. I can't do it alone'" (White III 1984, pg. 62).

Dr. Davis (formerly Williams) is now a professor at Florida Agricultural and Mechanical University (FAMU). A write up of Dr. Davis's outreach and educational work appears in the publication *The Capital Outlook*, an online publication that describes itself as "…the leading publication for mature and forward-thinking African Americans."[2] In addition to receiving a Master's of Business Administration in Marketing from the University of Texas, she pursued and received a Doctorate in Education, specifically in Science and Research Education, from Baylor University. In 2015 she published a book about educational theory in STEM entitled *How We Really Learn: The Micro-Spiral Methodology*.

9.4 The Importance of Sharing Narratives of Struggle and Moving Away from the Positive Image in Oil

By the 1980s the optimism of petroleum started to wane with the so-called "Oil Glut" that occurred throughout that decade, which caused the demand for oil to drop (Basosi et al. 2018; Jacobs 2016). In the 1986 edition of the magazine *Savy*, with photographs appearing in Fig. 9.7, Kathleen Fitzpatrick wrote an article entitled "It Happened to Me: Rock Bottom," in which Fitzpatrick experienced job insecurity because of the lack of demand for petroleum (Fitzpatrick 1986, pg. 18–19).

Fig. 9.7 Fitzpatrick and the eye-catching headline. (Fitzpatrick 1986, pg. 18–19)

[2] See http://capitaloutlook.com/site/about-us/

Fitzpatrick graduated from the University of Georgia, from which she graduated into the problematic oil market: "My graduation from the University of Georgia had coincided with the beginnings of the oil-boom-gone-bust and oil company merger-mania" (Fitzpatrick 1986, pg. 18–19). She had a dilemma with one of her first jobs: either transfer or get laid off after the merger of her employer, Cities Service Company, with Occidental Petroleum. She had been working in Tulsa, Oklahoma in geology, but then began a new job in Eunice, New Mexico, in the oil field. Her position was informally entitled "roustabout," which refers to a worker who does lots of different jobs (Fitzpatrick 1986, pg. 18–19). Fitzpatrick described the demands of the job, "By day I labored with wrenches, shovels, pipes—and dominoes at lunchtime; I spent my evenings sending resumes to oil companies."

She was able to land a geology job in Oklahoma City at MCO Resources, Inc. The year 1984 looked brighter for Fitzpatrick, as she was working in her field during the turbulent times of the 1980s, which she referred to industry destabilization as the "shakeout." She was also listened to at the company, as the business followed many of her recommendations and she was later rewarded with a 12% salary increase. She also received recognition from her boss through a memo. She was feeling professionally confident and also serving the field of geology, editing the newsletter of the Association for Women Geoscientists, as she had a chapter that she was active in and the group was easy to access. She also published in the *AAPG Explorer*, which was put out by the American Association of Petroleum Geologists. However, things could change very easily in the 1980s in petroleum geology.

Fitzpatrick looked at a condo in Oklahoma City and ultimately bought it. She was able to purchase it for a mortgage. Upon returning to work, she was laid off. Her boss summoned her, and everyone else on staff, to his office and explained the situation of the industry: "As you know; the oil industry is depressed now, and a lot of companies, including ours, have had problems. The decision has been made to close the Los Angeles and Oklahoma City offices on August 1. Most of you will lose your jobs..." (Fitzpatrick 1986, pg. 18–19). Fitzpatrick would soon be without a job, but still had a mortgage on her condo. Again, Fitzpatrick was faced with the harsh realities of a struggling job market and stuck with the condo. By August she was collecting unemployment paychecks. However, she was able to pick up some contract work, but it was low paying.

When she began to celebrate having her home and attempts at cheap living, she was able to secure a job offer as a geologist in Albany, New York, and return to the Northeast she had come from. She took the job. She was able to work out a deal with the bank where she turned over the deed in a "deed-in-lieu-of-foreclosure" and got rid of the condo without damaging her personal finances. Upon writing the article, she was working at the Bureau of Oil & Gas Regulation, in the Mineral Resources division in New York. Fitzpatrick ended the article with humorous acceptance:

> I am sitting in my rented apartment 100 miles north of my high school, where in tenth grade I chose to take earth science and geology instead of biology, because I didn't want to dissect fetal pigs. There were several times last year when I considered dissection a preferred occupation to geology. I also considered a new career in comedy. Did you hear the one about the

> geologist who bought a condo? A funny thing happened on the way home from closing… (Fitzpatrick 1986, pg. 18–19)

Fitzpatrick's experience was likely representative of many geologists in the petroleum field in the 1980s.

On September 16, 1984, *The Michigan Daily* ran a story examining the experiences of women in the sciences entitled "Program Guides Women into Science" (Roth 1984). The subject of the story, Catherine Badgley, was a graduate of the University of Michigan and was at the time of the publication of the article, a professor at the University of Michigan. Marking sixteen years since starting at the University, and after a fourteen-year stretch, she now was a working paleontologist. The article praised Badgley's work in the field of paleontology and as a woman: "Badgley broke into a male-dominated profession, but said she doesn't view being a woman as a hinderance to her career" (Roth 1984).

Badgley saw her own identity without the primary category of gender. She proclaimed that "…I see myself as a scientist first, then a woman" (Roth 1984). That year, the newspaper reported, she was participating in a workshop about women in science. The workshop was supposed to promote the role of women in science, and the article reported that most female scientists did not think gender was a really important issue in their scientific education, "Though some woman still complain of feeling shadowed by the larger number of men in science classes and of extra attention some professors give their male classmates, most agree that gender is not an issue in their academic success."

Students, who were featured in the article, reported and speculated on the identity and scientific education. Entries from students were included by the article's author, Marlene Roth, and were anonymous. The article made the decision to take the perspective of a male student. One entry included and featured an LSA student who was later identified as a man, as using the name "he." The quotations include that "'YOU MENTION women in science, I don't view them as quote 'women chem. Majors,' said one LSA student who asked not to be named" (Roth 1984). Additional quotations from the student continued that "'Yeah, there aren't as many girls in my (physical) chemistry class, as there were in my 125 lectures freshman year, but it seems almost archaic when you start separating the sexes,' he said." The LSA student continued that "'Hey, I'm a guy and you're a girl, but in class, we're working for a similar goal and then we're students,' he added" (Roth 1984). Other students expressed optimism for women in science on the job market, as they felt that there would a lot of companies that would want women: "A biology major said if the attitude that gender doesn't matter 'carries over into the job market, it will open up a lot of positions for women'" (Roth 1984). The workshop included many women from disciplines at the University, including the medical school, microbiology, and bioengineering.

9.5 "Women Learning to Survive" and Taking Their Place in Management

Women were entering the workforce after the 1960s in larger and larger numbers (Aig 1982). *The Nevada Daily Mail* reported in 1982 that the Census Bureau had 47 million women working and 3.1 million in that number were managers (Aig 1982). The article featured women who were managers and who also struggled with organizing child care and taking time off of work in order to recover from childbirth. The article featured the experiences of Pam Newman, who was one of the executive directors of Marsh & McLennan, an advisory business located in New York. She was almost at the end of her pregnancy term and the article author recorded that Newman was excited about her shortly arriving child "I like having children…But traveling is essential to my job" (Aig 1982). Newman reported in her personal history that after having her first child she returned quickly back to work, as quickly as two weeks later.

The reporter historicized that women, just a generation ago, would not have had as many opportunities as Newman did in her current job. The author, Marleene Aig, contextualized Newman's life: "Twenty years ago, it's unlikely Pam Newman would have gone so far – in her travels or her career. Even 10 years ago, she might have been expected to stay home to care for her children" (Aig 1982). The article also discussed a previous blockbuster book, *The Organization Man*, is no longer useful in discussing the workplace of the 1980s (Whyte Jr. 1956). *The Organization Man* described the postwar decadence and opportunism that businesses held out for aspirational middle-class white men. The book itself a critique of such businesses (Hanson and O'Donohue 2010).[3] Aig wrote that the 1980s were different, "More women are joining the corporate world to dictate letters instead of just taking them." She also historicized why more and more women were entering into the workforce:

> Women's liberation, civil rights legislation and economic necessity—from two-paycheck families to divorced mothers—have created this new breed of women. Armed with credentials and ambition, they are saying that being woman, a wife and a mother is no reason to abandon dreams of reaching the top. (Aig 1982)

She then compared how two corporations had changed in regards to the gender of their management. Xerox, a company that was in the mimeograph, and later electronic copier business, had around one woman for every male member of management in the 1970s. However, only ten years later there was one-woman manager for every seven male managers. The previous year, they had an increase of seven hundred female managers. Many of these women also held Master of Business Administration graduate degrees (MBA). The author found that women now make up twenty percent of graduate business programs that offered the MBA, while fifteen years ago there was hardly any female MBA students. Gulf Oil Corporation

---

[3] University of Pennsylvania Press has recently reissued the book: https://www.pennpress.org/9780812218190/the-organization-man/

found that there were fifteen hundred female owned businesses, and they published a guide of those businesses.

The article went on to discuss the accommodations that businesses were making in order to retain female talent (Aig 1982). These accommodations included day care centers, and corporations are taking deductions on their yearly taxes for providing childcare for employees. These child care centers are also subsidies by the government under the Economic Recovery Act of 1981. The article noted the effectiveness of the childcare centers, as these centers were credited with retaining employees.

However, there were problems for women, especially in their earnings. The article criticized the inequality of salary between men and women: "Women still climb the corporate ladder more slowly and for less money than men. Salary statistics tell part of the story: Only 0.8 percent of full-time working women earn $25,000, compared to 12 percent of such men" (Aig 1982). In comparison of inflation as of 2023, $25,000 would be roughly equal to the buying power of about $76,000 (Aig 1982).[4]

There were many reasons as to why women were not reaching management positions. Older studies performed by Harvard Business School were cited in the article, stating that around half of the male executives surveyed found that women were not able to hold positions of leadership. But more recent literature cited in the article found that women did not reach higher level management positions because they did not get mentorship and constructive criticism, compared to their male peers. There was an asymmetry of knowledge between women and men in knowing how to corporate and utilize resources in corporations. Margaret Henning, a Dean of the Simmons College and professor of management, discussed her college, and others, offering courses that helped women navigate the male-controlled business culture. Other experts in management and business advocated for the continuation and usefulness of women's skills in management positions.

In 1966, Adele S. Greenfield was one of the only women who served as Chief Executive Officer in the American oil industry (Alama 1966). An image of Greenfield appears in Fig. 9.8. One of the subheadings included with the image of Greenfield is entitled "Pulchritude in a Man's World." According to the *Cambridge Dictionary*, pulchritude usually refers to an intense feminine beauty. Though Greenfield had a high-level administrative position, profiles still utilize beauty to define the working woman in petroleum. The image captioned also had her talking to a male author who wrote about competition at the Empire State Petroleum and Fuel Merchants Association meeting in Syracuse, New York. The article emphasized that beautiful women were working successfully in a very competitive field.

The article opened with the emphasis on the beautiful woman in a highly competitive area: "A WOMAN can find a place in the competitive world of the fuel oil business and continue to retain her femininity." (Alama 1966) Greenfield, who was named as the "executive director of Oil Heat Institute of Westchester County," says that a "woman's place depends on one factor[,]" which she emphasized as

[4] This analysis was estimated through a variety of online electronic calculators.

Fig. 9.8 Adele S. Greenfield. (Alama 1966)

"competence." She emphasized that the ability to do one's job well does not mean sacrificing her femininity: "Competence, she says, and not femininity is the stand of measurement in her business. She knows. She's the only woman executive in the business in the country" (Alama 1966). Greenfield was awarded for her excellence at her job by the Empire State Petroleum and Fuel Merchants Association. She stated that she performed her job to such a high standard that men treat her equally: "'Men accept me and treat me as an equal,' declares Mrs. Greenfield who is the wife of Norman Greenfield, a wholesale meat producer."

Adele Greenfield worked her way up into the executive job, as she had been fired from one job, and sought a secretarial job at the oil company. Her previous employer, her husband, had fired her from her job after what the article records as a "dispute" (Alama 1966). Like the World War II and early postwar profiles of women working in oil, her family, community, and volunteer work was emphasized in describing her accomplishments. It was after her children, as she described reached "maturity," she returned to the workforce and her first job was with her husband. The article inquired if there were any advice about how to be as successful as Greenfield, and she responded that she treated other men as if they were her spouse. And like the profiles in the previous World War II and postwar eras, her physicality was showcased, but the profile emphasized how she was powerful in demanding equality: "And the men, in turn, respect her for her ability and knowledge of the position, the 5 foot-3 inch brunette (5-5 in heels) says with aplomb" (Alama 1966). Her competence was continually stressed in the article "…she's prepared to tackle any type of job." And like the profiles of June King, she participated in all aspects of the business: "In times past, Mrs. Greenfield rode oil barges, accompanied fuel oil distributors on their trucks. She methodically keeps the membership up to date on all national, state and local legislation which affects them" (Alama 1966).

Greenfield emphasized the pride she received from demanding equality in the workplace. She acknowledged that, "'Men help [me] and give me the greatest compliment of all when they accept me professionally. There is a definite place for women in this business if they sincerely want to learn about it and devote their efforts to it,' Mrs. Greenfield believes" (Alama 1966).[5] She also emphasized that competence trumped both attractiveness and sexism, "'Men in business are more concerned about performance than about feminine wiles,' she declares" (Alama 1966). However, the writer of the article, a man by the name of Malcom R. Alama ends the article on a questionable note by adding a line: "Want to bet?" and the meaning is unclear. The ending either means that the writer was questioning whether sexism was still be a persistent issue in the workforce for female CEOS, or it was a sexist swipe about women needing to maintain femininity in order to command equality or competence at their job.

Women seemed to see opportunities in management opening up, but were challenged by the sexist ideas of womanhood and femininity as being necessary components to proper leadership. Profiles of women, even in management, continued to run glamour profiles but even those that emphasized sexism, discussed equality, or the desire for equality in women performing their jobs as leaders. But profiles like those written about Greenfield, with historical context, seem to question the relationship between glamour and the demand for equality in the workforce in the 1960s.

9.6 Discrimination, Economics, and the New Glamour

In 1950, newspapers like *The Altus Times-Democrat*, published out of Oklahoma, not only covered news in the oil industry but also published an article that explored the pressure and anxiety of the working women in the oil industry. The story, out of Norman, Oklahoma, remarked that "Women are competing successfully in many fields once considered exclusive professions for men, six University of Oklahoma educator[s] believe" (The Altus Times-Democrat 1950).[6] But those same educators also remarked that women have to work harder and more diligently than men in those same jobs. And the pressure of the job market was spilling into school culture and making for a pressure-filled environment for female students: "Good grades in college were stressed [sic] essential for women completing in male fields. Wartime [afforded?] women opportunities to prove their abilities, the Sooner professor agreed" (The Altus Times-Democrat 1950). Many pink-collar jobs were highlighted, or those jobs that sociologists categorized as pink-collar because were often

[5] In the original quotation contains the phrase "Men help we" and I changed we to me to make it more readable (Alama 1966).

[6] The newspaper article appeared faded and was difficult to read and included many missing words. The author did the best that they could to read the article (The Altus Times-Democrat 1950).

female dominated (The Altus Times-Democrat 1950). And the professor interviewed was optimistic about the potential of women taking positions in other fields. They were especially optimistic about women entering into the petroleum industry, with a professor remarking that, "'Women graduates of OU are in the petroleum geology field from Canada to Texas,'…'One out of 25 geology students is a woman and they have no trouble finding jobs" (The Altus Times-Democrat 1950).

Women also received negative statistics about the income gap that continued to widen in the 1970s. The Labor Department's Women's Bureau that was involved with making promotional materials to recruit women into geology and petroleum geology reported that they found a "grim picture," in which as of 1970 women were making 57 cents of a dollar compared to men, and it was even lower than the earnings in the 1950s, when in 1955 women made 64 cents of every dollar compared to men (Lyons 1972; The Altus Times-Democrat 1950). The scientists who determined this statistic, a Mrs. Elizabeth Duncan Koontz, reported it at the meeting of the American Association for the Advancement of Science meeting.

The panel at the AAAS meeting focused on working women and their feelings of discrimination. Elizabeth Duncan Kootnz was an African-American educator, National Education Association Leader, and leader of the Women's Labor Bureau (Lyons 1972; The Altus Times-Democrat 1950; Alexander 2022). One theme was the inequality, and also inability, for men to share familiar powers, which the article summarized as the woman's panel: "Their theme was that men are afraid to share social and familiar power and responsibilities with women, and by not doing so, only work themselves into their graves by carrying either most or all of the financial burdens." (Lyons 1972).

Koontz presented a paper, that despite searching, I was unable to find, but was entitled "Myth and Reality in the Employment of Women" that was on the program of AAAS in 1972 (Lyons 1972). In the summary of her paper, and other presenting at AAAS, she found that only seven percent of women made 10,000, compared to that of forty percent of men, and in 1972 $10,000 was a lot of money, with the purchasing power in 2023 equal to roughly $70,000 dollars (Lyons 1972). She found that women only made up of around ten percent of professions like law, engineering, and medicine, while around ninety percent of women made up nurses and teachers. Other women on the AAAS pattern worried that women were having their potential squandered, while other members disputed biological differences that were said to explain different employment patterns. One panelist argued that this labor pattern was physically dangerous for men and often led to high incidences of heart attacks, and questioned the audience with, "Are you willing to keep your wife at home and kill yourself doing it?" (Lyons 1972).

In 1970, Koontz expressed optimism, "DURING the past 50 years, women have achieved greater freedom and opportunities, but they are still not recognized truly equal with men in many facets of our society" (Koontz 1970; pg. 3). She added that women were growing increasingly frustrated by their lack of equality, "Their dissatisfaction has increased and the ranks of women's 'liberation' movement, which goes by many names, are growing" (Koontz 1970; pg. 3). Koontz framed the new women's movement, which went beyond advocating for the right to vote: "Today, the movement is attacking the status quo through varied legal and social channels, and again dissident women are taking their grievances to picket lines, protest rallies, and marches on Congress" (Koontz 1970; pg. 3). Women were focused on taking more equality for themselves, "It is clear that women's renewed discontent with a secondary position in our society and their urgent demands for reforms will direct future activities for the Bureau toward new areas of concern" (Koontz 1970; pg. 3). Taking their place in society and becoming more equal meant fighting against discrimination, "Women in various situations—as workers, as students, as members of the community—and women at all economic levels face discrimination and are concerned, not only for themselves, as individuals, but for all women." It was the Women's Bureau's "responsibility" to help them fight against discrimination, "…to coordinate those concerns to help women recognize and utilize the means available to them so that they can make an orderly, rather than a disruptive, approach to the problems of discrimination" (Koontz 1970; pg. 3). She was focused on working to fight discrimination for women at the workplace.

Women also faced outright discrimination in the workplace, but were recorded as trying to spin the story to minimize the challenges of their workplace. One example can be found in the article about Bonnie Butler Bunning. Bunning was portrayed in the same type of attention-grabbing article format of the postwar era, but the 1976 profile was different in that it highlighted the discrimination and struggle of women in geology with the headline "Geologist Notes Bias" (Davis 1976). At the time of publication, she was 25 years old and she had worked in mining in Arizona and had superior qualification. The article discussed her reasons to proceed into industrial geology, but also emphasized the challenges of balancing a dual family household, "One [Bunning] of a very qualified female exploration geologists in the Inland Empire, Ms. Bunning said her decision to be an 'economic geologist' was made in concert with her journalist husband's decision to take a newspaper job in Spokane" (Davis 1976). Bunning looked for work after her husband received his job at a Spokane paper, which might have been the *Spokane Daily Chronicle* that published this 1976 story, complete with her image in Fig. 9.9.

Bunning acknowledged the challenges that were summarized by the president of the Northwest Mining association in that there was historic sexism in hiring that Bunning agreed that there was: "a residue of prejudice against women in mining field work[.]" (Davis 1976). She was the first female applicant her hiring manager had interviewed. Bunning alluded to the inexperience that her hiring manager had with women, and recounted her experience: "It must have been really 'wow' for him or something because he told me there would be a lot of hiking and strenuous field work and that I probably couldn't handle it[.]" (Davis 1976). She had a similar experience with other men who interviewed her for mining positions, even citing the attitude of the company against hiring women, "He doubted [the additional male interviewer] the company would hire a woman because "it was pretty well known that women party all the time" (Davis 1976). Bunning then responded that there was a type of woman in the field who was interested in romantic pursuits, adding to the summary saying that, "'You know the type,' she said, 'they nudge each other, wink

Fig. 9.9 Butler at work with maps. (Davis 1976)

and say, 'We all know what happens when a man and woman go out in the field together,' that type" (Davis 1976).

Though the president of the Northwest Mining Association convention acknowledges that there was still sexism in the field, there was little done about it. Bunning recounted her inability to fully participate in professional societies like the Northwest Mining Association, as she was refunded for her full membership while attending a conference because women could only be associated members. Bunning was understandably upset by the whole situation, "'I don't even remember how it all turned out. It was ridiculous. I just wanted to attend the convention as a practicing [space] exploration geologist for Homestake Mining Co. The Company paid the fee,' said Ms. Bunning" (Davis 1976). The article later reported that this "incident" should not have happened and women were changed to full membership, and a "representative" stated that this was not in keeping with the organization's practices.

Bunning praised her education at Smith College because she said that there were both no limits there or "competitions," and she found that this was unique to Smith compared to larger educational institutions, and she felt that she could pursue additional or graduate education later (Davis 1976). She also valued the environment and nature, "She describes herself as 'realistic environmentalist noting that no mining venture which could leave scars on the earth should be started unless planners have reclamation in mind at the outset'" (Davis 1976). She felt that protected recreation areas were insufficient to protect nature.

At the same time that women were fighting against discrimination in the workplace and in science, newspaper articles framed the lack of women entering into science was a societal problem (Myers 1960). The lack of female participation could be seen at the meetings of the American Association for the Advancement where most of the papers were read by men (Myers 1960). In 1958, Myers recalled, less than five percent of women had college degrees in science, and ten percent of men held science degrees. In 1958, of the 170,000 scientists in America, there were only 14,000 women, and most of those women were chemists. Myers referred to the work of Alice Leopold, whose work was explored in chapter three regarding careers for women. Myers called out the "Folklore" that math and science were "not for girls" (Myers 1960). Sometimes high school guidance counselors "…steer scientifically talented girls away from science and mathematics" (Myers 1960). Myers cited Leopold's advice that better scientific preparation in high school will lead to more scientific majors.

Women's colleges seemed to help with the problem of women not entering scientific education. An article that covered the marketing of women's colleges in meeting the needs of women stated that it was only women's colleges that could understand the unique problems women experience in learning: "Many women's colleges have pioneered programs relating to women's needs. These include new courses introduced into the curriculums as well as entire institutes…" Other women's colleges in cities have made their scheduling more "flexible" for working women and mothers, as well as adjusting their required classes for older women returning to school (Lewiston Daily Sun 1983). Even though Title IX was law in 1972, the Association of American Colleges found that "Women were not being encouraged to prepare themselves for careers in non-traditional fields, such as math and science. The discrimination was subtle but nonetheless present" (Lewiston Daily Sun 1983). Students found that attending women's colleges had better, noncompetitive climate in the classroom, students felt empowered, and as one student was quoted as saying that "You're not intimidated by guys in the class. You can get more out of your studies." Students also stated that better friendships were created on campus. Women's colleges seemed to be an incubator in the history of women's scientific education (Tolley 2003).

The internalized pressure was also noted at the end of the article explaining the experiences of Bunning. Bunning felt pressure because of her status as a woman, and she felt that she had to excel in her profession: "'There is also a lot of self-imposed pressure to do well because I am a woman and because my effort might play a role in the success of women who might apply tomorrow for a job with a local mining company,' she said" (Davis 1976). But she was accepted by her immediate colleagues, but not the industry in general, "According to Ms. Bunning, her colleagues accept her as a geologist but that general industry acceptance of women in mining will take some time" (Davis 1976). She also saw herself in a mentor type role, which was a real break from some of the women in the 1950s, such as Kitchen, a geologist mentioned in the previous chapter, who was never described by her first name. Bunning saw herself as needing to succeed to help all present and future women succeed, "'I try to encourage women geology students. Things are getting better but I don't see the day of women on offshore oil rigs happening very soon. That day is a long way off but it is coming,' she forecasts" (Davis 1976).

Geologists in other related fields to petroleum geology continued to put a positive attitude towards the sexism and ageism that they experienced at the worksite. One example comes from North Carolina, where Sharon Gardner (who was named Sharon Lewis prior to her marriage), worked as the first female geologist for the North Carolina State Highway Commission. Gardner worked with drilling machines in order to find suitable road sites in North Carolina. Locals were often caught off guard at a working woman geologist in 1972, as the article shared the shock of a local man, "An elderly local man chatting with two members of her geologists party, rocked back on his heels, blinked his eyes and shouted out in astonishment, 'You mean you've got girls working for the Highway Commission now?'" (Merdecal 1972). The response that Gardner gave the older local man was just a smile. Gardner's image, as a working geologist, appears in Fig. 9.10.

Gardner was described as twenty-five years old and "soft-spoken young woman" who had become used to defying expectations. She was one of the first women to instruct laboratory courses at Vanderbilt, where she received both her undergraduate

Fig. 9.10 Gardner appearing near equipment in the field. (Merdecal 1972)

and graduate education. But Gardner remarked that women working in geology had to work against a lot assumptions by society at large: "Frankly, it's harder for a woman geologist to get a job…It takes a lot of getting used to, to see a woman in the field with a hard hat on using a drill machine" (Merdecal 1972). Women like Gardner had to use self-deprecating humor to work through the sexism and difficult workplace that existed for working women: "In her branch of the Highway Commission in Raleigh, Mrs. Gardner is the most recently employed of five geologists. She jokingly calls herself the 'general flunky' of the office and her opportunity to work in the field is limited" (Merdecal 1972).

The description of her appearance focused on her working clothes that did not set her apart from other working field geologists. Gardner was described as wearing hard hats, "dirty jeans," instead of clothing like a dress. When she was away from the field, she was busy in the office drafting out maps. The article also acknowledged that geology was becoming less popular for women. There was a lot of experiences and episodes that might have led Gardner away from geology for continuing her studies but she continued to literally break ground.

The article historicized Gardner's experiences, "Sharon Gardner is breaking ground in more than one way in her profession. Geology has been a less than popular field for women in the past. Those few who were interested were discouraged by more traditional professors" (Merdecal 1972). One professor at Vanderbilt would not allow women to attend his course because he did not want to moderate his behavior: "When Mrs. Gardner attended Vanderbilt, the head of the geology department barred women from his own advance classes. 'He was a crusty old character who wouldn't let women in because he use foul language and didn't want to tone it down for them,' she laughed." Again, Gardner tried to use humor to mitigate the problems in geological education.

Gardner, despite the challenges in her education at Vanderbilt, found a real love for geology. She had a long history of interest in rocks while she grew up near the Appalachian Mountains in the state of Kentucky. She came from a family of coal miners, and her father would bring her fossils and other rocks of interests that he found on the job site. She also grew up in a mining community that still believed that women were bad luck at mining sites. Her husband was also a working geologist, whom she often discussed her work with after the end of the day.

Gardner was not alone in her experiences. Brenda Higgins was the first woman hired by the state of Florida's Department of Transportation (Ocala Star-Banner 1970). Higgins profile ran in additional newspapers in Florida (Sarasota Herald-Tribune 1970). Higgins was not optimistic about women reaching parity with men in science. The article that ran in the 1970 *Ocala Star Banner* began the article by the question whether "Full equality for men and women?" Brenda Higgs was pessimistic even though Higgins was one of the first women to be hired by the Florida DOT as a geologist: "It will never happen, says Brenda Higgins, who is the first female geologist ever hired by the Florida Department of Transportation (Although Brenda is the first female geologist in the Department, DOT presently has 19 women filling [sic] profession positions)" (Ocala Star-Banner 1970). But she argued that the women's rights movement allowed her to have a good job at a proper pay rate.

The article contextualized in bold font that, "For the first time in her career, Mrs. Higgins is receiving 'equal pay for equal work' and she feels this platform is the most worthy espoused by the Women's Liberation Movement" (Ocala Star-Banner 1970).

She still remained pessimistic about her role in the working world and women's opportunities to make money truly equal to her colleagues, arguing that, "If I weren't so apathetic, I'd join. I think many of their goals are worthwhile. I don't see why men should receive higher pay than women. Look at secretaries, they perform work equally as difficult as many jobs performed by men. Invariably their pay is lower" (Ocala Star-Banner 1970). Women like Higgins did not see the value in the special treatment that women receive in compensation for their lack of equality and would much prefer having fully equality with men: "Women, thinks Mrs. Higgins, wouldn't be giving up much to archive quality. 'We've only received special consideration in little things—like having [sic] a cigarette lighted. Actually it's easier to light your own'" (Ocala Star-Banner 1970). She did not see the Women's Liberation Movement (WLM) as ever achieving the goal of equality, and she acknowledged that the only real change is the current change of women receiving more equality, but not total equality.

Play was discussed in her development of interest in geologist, as Higgins rejected traditionally feminine toys of dolls. The article argued that she was interested in scientific collecting from an early age: "While other little girls were playing with their dolls or tea sets, she was collecting plant fossils in the coal mining areas of Western Pennsylvania" (Ocala Star-Banner 1970). After some time in universities in Pennsylvania she got both her undergraduate and graduate education at the University of Florida. She had one year in the Ph.D. program there in Metallurgical and Materials Engineering, where there were mostly male students. After some time in research, she eventually got a job with Florida DOT, but it was not without experiencing difficulties in a mostly male profession.

The article described her difficulties on the job market but also minimized her experiences with discrimination: "Although she encountered masculine prejudice against women in 'a man's profession only once [sic] in college Mrs. Higgins met with difficulties in seeking employment receiving her degrees" (Ocala Star-Banner 1970). The article, however, recorded, more detailed of discrimination that Higgins experienced while pursuing employment, "Despite de[creasing] prejudice against women in the scientific fields, Mrs. Higgins said women applicants for geology posts were often given the standard excuse 'Much of our work is in the field and you can't send a woman into the field[.]'" (Ocala Star-Banner 1970). She recounted the most overt discrimination that she had experienced on the job market when she applied for an unnamed government position as her interviewer explained that, "One of our jobs is probing the river. First you'd have to get a boat off a car, then put a motor on the boat and stretch a steel cable across the river. Then you'd have to put 100 pound weight on the cable. And if you could do the rest you couldn't lift 100 pounds in weights" (Ocala Star-Banner 1970). The article asserted that Higgins got not only fair pay compared to men, but proper treatment by male co-workers. A picture was included to emphasize the collegiality of Higgins and her male colleagues in Fig. 9.11.

The article also emphasized that Higgins was equal to her male colleagues in regard to the performance of her job. The article described her duties at DOT, "And well she might, since she works with her masculine counterparts doing [sic] X-ray diffraction studies, making petrographic examinations, elemental analyses utilizing the electron microprobe, and making field trips to investigate bridges, sinkholes and other areas requiring geological examinations" (Ocala Star-Banner 1970). The article ran images of Higgins and seated with her co-workers, which appear in Fig. 9.11.

The article ended with a question in regard to "How does THE man in Brenda's life feel about her career?" In the 1970s, it was becoming more common for relationships to include both members working (Seiler 2008). Her description lacked any friction from her working in geology, "There was no career conflicts, says Brenda, although she occasionally balks at stopping the car to permit investigation of the unusual rock outcrop" (Ocala Star-Banner 1970). And the article promoted the relationship as they shared interests and seemed to have a positive relationship, "He and his wife share many common interests. Both are sports car buffs and Brenda says of husband Alton, owner of an electronics business which specialized in biomedical equipment, 'if you ever want anything mechanical explained to you, he's the one who can do it'" (Ocala Star-Banner 1970). Profiles in the 1970s of women did emphasize discrimination and the trials of women working in geology that still retained sexism, these profiles emphasized aspects where women were working through the sexism and making a good living, even having equality with their

DOT'S BRENDA—Looking more like a college coed than the experienced geologist she is, Brenda Higgins checks a map in her duties as geologist for the Florida Department of Transportation (DOT). In photo below, she poses with fellow geologists Steve Denahan, Paul Beam and Bill Wisner.

Fig. 9.11 Higgins pictured at work with maps and with colleagues. (Ocala Star-Banner 1970)

romantic partners. Having professional and personal success became a trope in these articles, as the dual earning household produced material wealth (Ocala Star-Banner 1970).

Calls for women to enter engineering and geology remained in the optimistic feelings of the 1970s, and hiring managers had difficulty filling the demand for positions (Link 1973). In 1976, the story based in Washington, proclaimed that women engineers were in high demand from industry: "'Engineering Elites' are the women most in demand by American industry, reports Industry Week magazine in a recent survey" (Link 1973). Companies like General Motors were increasing the number of women in management, but women studying engineering in college was smaller than one percent of the overall college graduates in engineering majors. American Rockwell Corporation wanted to have more women working in the new space shuttle division. Larger Corporation leaders, like a vice-president out of Eaton, complained that women's education underserved the needs of companies: "Most women in their twenties and thirties simply are not trained properly. Their educational background is mostly in the arts and that's not geared to the needs of industry" (Link 1973). While other corporate spokesmen stated that women wanted to work more directly with the customer and not in "heavy industry" (Link 1973).

Industrial researchers claimed that another problem of women entering a heavy industry like petroleum came from the problem of image and imagination. They stated that women had a hard time visualizing themselves working in petroleum, "It is difficult for women to see themselves in the petroleum industry, with its strongly masculine image, and for men to see that masculine image changed" (Link 1973). The lack of women in petroleum engineering hurt both male and female workers. Combating this image problem likely continued the marketing efforts of oil companies of profiling women working in geology as a glamorous job, but one not without difficulties.

The article continued to record problems related to women in the workforce, but many of the additional reasons given by industrial spokesperson continued to blame women for their circumstances. Dow Chemical claimed that women were simply not "aggressive" in pursuing employment and planning for employment. Dow noted that women had irregular career "patterns" where they left and re-entered the work force based on familiar obligations, like marriage and children. The article concluded by inquiring about equality in the work force, but industrial leaders were agnostic about their companies reaching equality goals,

> While most of the companies surveyed by Industry Week agree that government agencies are "demanding" in their follow-up checks on compliance with equal employment goals, one executive wondered if those goals were realistic. "If we say we're going to balance our cross section of employees so that it matches that of the population, as we do with equal employment goals, one executive wondered if those goals were relations. "If we say we're going to balance our cross section of employees so that it matches that of the population, as we do with blacks, do we want 40 per cent to 50 percent of our jobs filled with women? After all, what about the large number of women who want to be "sex objects." (Link 1973)

The 1973 article was written around the time in the same decade as the Higgins profile, and readers of this book can certainly understand the pessimism and frustration of women working in geology with executives having that type of mindset in hiring practices.

Women also worked to fight discrimination at their jobs through labor organizations and unions. *The Eugene Register-Guard* ran a story about a working geologist, Nellie Fox, who was one of the first women to lead the AFL-CIO in Oregon, and had career aspirations to become a geologist. She attributed her sex as preventing her from achieving her career dreams: "After she graduates from Franklin High School in Portland a wealthy relative told her, 'If you had been a boy, I would have sent you to college.' So she didn't become a geologist" (Eugene Register-Guard 1976). She started a family and got a job making one dollar and twenty five cents as a secretary. Fox noted that her job had caused disagreements in the home and she had a pattern of stopping and reentering the workforce. She had to continue to work after her husband, who had encouraged her to get into labor advocacy, got ill with a "rheumatoid disorder" (Eugene Register-Guard 1976).

After returning to retail merchandising sales, she entered her Local No., 1257 as a representative of her labor union, but also received pushback for being active in the labor movement, as she was told women were supposed to remain in the home. She expressed her experiences being discriminated as a woman in the workforce and in organized labor, "'It wasn't ladylike for women to belong to the labor movement in those days…'And there was the fear that working women would take jobs [sic] away from men" (Eugene Register-Guard 1976). She argued that women were blamed for taking away jobs from men. Women unemployment numbers had started to climb, but this is not becoming as large a problem, because women were entering into organized labor union governance. She argued that women became interested further in the labor movement with the rise of African-Americans entering the workforce: "Women are becoming more active and more are being elected to office. 'It was only after blacks started opening doors, that women really become involved'" (Eugene Register-Guard 1976). Fox said women had a history of arguing for better pay and ending discrimination. She advocated the women's movement to women who might be against it because it projects them economically. Fox argued that, "To the women who downgrade the feminist movement, she says 'They're only a heart attack or a divorce away from supporting the movement. They learn in a hurry it they're faced with some of these problems'" (Eugene Register-Guard 1976).

Fox found that some women have nicknamed themselves "until workers" as they will become more vocal in their support of feminism when their major bills are paid, such as their children are sent through college or their mortgages have come to an end. But Fox argued that they were getting lower wages and younger women needed to be aware of their rights, and schools should encourage young women to work in the "trades" as they provided a "realistic" living. Fox remained interested in geology, in addition to her work in organized labor. She described herself as a "rockhound" but she had to leave her hobby aside temporarily as she had to move to the capital of Oregon to work for labor rights. The article ended with the acknowledgement that her husband will be taking care of her two dogs and cats while she was traveling for work.

9.7 Personal and Professional Discrimination in the 1970s

Even historians and practitioners writing about women in the petroleum business acknowledged their own personal experiences of discrimination. Robbie Gries in her book that is a very glamorous celebration of women in petroleum geology, *Anomalies*, described difficulties in her relationship that her working outside of the home (Gries 2017). She wrote that in her first marriage she attempted to be a stay-at-home wife and mother but found it unfulfilling, and she described her husband's inability to understand,

> I did get my degree, got married, and had a baby; but, I soon found I was not cut out for the housewife role. The daily, repetitions work maintain a household was frustrating to me after the fun and excitement of research and work in the UT geology department. Intellectually, especially being in a new town)(my husband I had moved from Austin to Wichita, Kansas) where I knew few people. I told my husband I just had to go to work ,at least part-time , and that I would use my salary to hire a maid to do the housework. He said, not once, but several times, "…if we got a maid, then I don't need a wife. (Gries 2017, pg. 250)

Gries and other women experienced difficulties transitioning into work in petroleum geology both professionally and personally, as in Gries own autobiography, she described pay discrimination and other difficulties in petroleum geology as she started working in the field of the time of Affiliative Action, which she described as requiring oil companies to hire more women.

Affirmative action has a long and complicated history in the United States and is still being examined and revised in education, business, politics, and the courts. Historians have bravely and competently studied the confusing and divided opinions and actions regarding Affirmative Action, and there is still a lot of historical work that needs to be done (Anderson 2004; Rubio 2001; (Urofsky 2020). In 1981, the US Commission on Civil Rights defined Affirmative action by three features. The first was that a person's race, biological sex, and national origin could be considered in order to assist in remedying past historical discrimination. Secondly, affirmative action is supposed to bring about more equitable opportunities. And finally, Affirmative Actions might provide for "protective class" that is aided by such actions (Urofsky 2020). Affirmative Action's purpose was to "overcome discrimination" and provide more equality to historically discriminated groups (Urofsky 2020). In colleges and universities, Affirmative Action was practiced, but there were court challenges. There has been lots of "trial and error" in the practices of Affirmative Action (Urofsky 2020). But scholars like Urofsky frame Affirmative Action as not a desired end result, it is supposed to be a tool. Some members of the Supreme Court assumed that there would be no need for Affirmative Action after a twenty-five-year period. The origins of the ideas associated with Affirmative Action might have started in the John F. Kennedy or Lyndon Johnson administration, or perhaps even in the Reconstruction Act after the Civil War (Urofsky 2020). Affirmative Action made a difference in the increase of the hiring of women in the oil industry (Gries 2017).

The 1970s had women working in geology and petroleum winning cases and large settlements on the basis of discrimination. Helen E. DeSanctis, based in the state of New York, and a working geologist at Shell Oil, won a case of discrimination against her age and her sex. She had been denied proper treatment at the company: "Mrs. DeSanctis, 51, had worked for Shell for 17 years when she charged last June that she was denied equal pay for the same work performed by men and was not given equal advancement opportunity" (The Toledo Blade 1973). She also missed out in advancement by being offered to "relocate or to take early retirement" when Shell moved its corporate headquarters from New York to Houston, Texas. The terms of her financial award made her ineligible to work at Shell.

In 1977, *The Evening News* out of New York City, published a profile of a female geologist working on an oil platform (The Evening News 1977). The tensions about women on oil rigs continued in the 1970s. Newspapers covered working women geologists working on oil drilling projects. But women chronicled in newspapers seem to omit many of the problems that were going on drilling platforms, and embraced the heroic and novel model of the female geologist, and these write-up included language of acceptance and narratives like those of June King.

Catherine Ann Ariey was described as working on an oil platform in Alaska as the only woman member of a seventy-five-person crew. Her work is described as an accomplishment and a counter-narrative as she "…has penetrated a traditionally male bastion of the oil industry and attained the willing acceptance of her peers" (The Evening News 1977). Her physical description was given as "the petite, 26-year-old brunette is responsible for analyzing and evaluation the oil bearing potential of soil cutting brought up from the drill bit turning beneath the water of South Cook Inlet" (The Evening News 1977). The vessel was constantly referred to using the language of exploring, giving the article a flair of adventure. But the article did also comment on her experiencing being a woman on a mostly male crew. The article casts Ariey's experiences, as does she, in a positive light, "At first the men tipped their hats to me, but after awhile the novelty wore off and I was accepted as one of the group. They call me Cass" (The Evening News 1977).

She embraced the opportunity to serve on the boat, as it was a great educational experience. She praised the time on the boat: "It's a great classroom out here…You need this experience to relate to your office work" (The Evening News 1977). Her quarters were also described as having her own bedroom and bathroom, as well as an office. She was able to recreate in the ship's facilities, and enjoy pool, movies, and exercise. Ariey's image appears in Fig. 9.12. She had been working for the Atlantic Richfield Company in 1975 and was working while she was finishing her Master's in geology from the University of Alaska. She had experience working in "wellsite geology" and living out of hotels, and her work as a geologist was quite involved "As part of her job responsibilities, she examined rock samples and monitored drilling mud for signs of petroleum vapors that indicated the presence of an oil or gas zone. In addition, she recorded her findings and reported them by phone to

CATHERINE ARIEY, WELLSITE GEOLOGIST
... on largest floating drilling vessel

Fig. 9.12 Catherine Ariey. (The Evening News 1977)

the Anchorage office" (The Evening News 1977). She had been pushing herself since her entry into the geology workforce in 1973.

Women like Cass were highly involved with field work in the remote parts of Alaska, where they collected microfossils for industrial surveys. They had adventurous experiences, such as riding by helicopter. Pursuing Cass's line of work was also portrayed as exciting, as she got to participate in her hobbies, such as skiing, scuba diving, and traveling across the outdoors. Her love of geology came from early child hood experiences in California at age eight, "That's when I began my rock collection[.]" (The Evening News 1977). She was also the lone female graduate student in geology at the University of Alaska, but there are additional women in the program.

The emphasis on defying assumptions and pathbreaking are still emphasized in profiles produced in 1977. Women geologists were profiled in many of the same ways that they were during World War II. The article emphasized the surprise that some people have when they heard where Cass Arley was working, and though she said her parents are very proud of her, it would be hard for them to comprehend her life on a sea oil rig. She explained that, "They refer to it like a battleship because that's what they relate to best..." (The Evening News 1977). The article noted that though there were other people of accomplishment in the family, she was the only person who had a picture on the family mantel with a "hardhat." She emphasized her plans to continue to improve her geologic knowledge through more field work, schooling, and study, and promoted the field to others, not just women, "Ms. Arley added that for others interested in geology she thinking a good broad background in the subject and lots of field work provide the basics for a good career. She said, 'The more rocks you see the better it is. Geological work can be hard but it's rewarding because you have an important job and the company is relying on your knowledge'" (The Evening News 1977).

Though there were accounts of Arely Cass about the positive work pace of ocean-based oil rigs, other articles discussed the problems related to anti-discrimination reform. An example can be found in the Scotland, especially the Sex Discrimination Act of 1975 (Cochrane 1979). Female oil geologists brought legal cases before the Equal Opportunities Commissions. The geologists were Miss. D Barrett and Miss J. Gray, who were employed by the British National Oil Company. The EOC was located in Glasgow, Scotland. The case was settled out of court and the women were able to pursue business opportunities on offshore rigs. Other cases brought by female scientists had been resolved out of the courtroom. The Sex Discrimination Act of 1975 was portrayed as a tool to encourage companies to settle discrimination claims: "A spokeswoman for the commission in London said time to time that there is discrimination reported to us but the oil companies tend to settle within the spirt of the Act" (Cochrane 1979). There were some periodic legal difficulties, such as the classifications of the rigs as boats, or whether the rigs had their flags of origins, but the article claimed that most companies, when suit was brought, settled and allowed the women to work. The article carried the reasoning behind industries not allowing women to work on rigs because of the lack of facilities and "…their presence might disturb the men" (Cochrane 1979). But oil companies had women working on rigs, such as Mobil Oil claiming that they had 12 women working, and other companies had women working in fields from geology, communications, and catering as well. But the article was entitled "Job Problem Over Women On Oil Rigs" which implied women were the problem (Cochrane 1979).

The cases of women not being able to work on oil rigs based out of Britain were examined further in the piece *Fear of the Frilly Nighties* that questioned "why are women barred from Britain's oilrigs?" (Coote 1979, pg. 206). The article opened with two female geologists who had worked on a deep sea well had their names removed from the duty "rota" (Coote 1979, pg. 206). The two "young geologists" were upset that they were unable to fulfill an "essential" part of their job and cited a summary of their job duties, "From time to time a geologist will be responsible for monitoring the drilling of a well at the wellsite, including continuous sample logging and use of gas detection equipment. He will also, in conjunction with the Drilling Supervisor, pick coring points and instruct the drilling personnel on the taking of cores and when the run wireline logs…" (Coote 1979, pg. 206). These duties could only be changed in emergency situations. One of the geologists, who noticed her omission off the rota, tried to contact her employer and asked why she had been removed, and she was informed that it was because of the disruption that she could cause.

The reasons for the removal of the female geologists from the deep sea well schedule were divided into four. The first was that men working on the rig might be seen without clothes, or insufficient clothes, and that a female employer might view them in an unclothed state. The second was that the presence of women on the rig might distract the male workers. The third was a potential objection from the wives of the male workers. The final was that the owners of the rig might not approve of women on the rig. The article reminded the reader that though these objections were expressed, female geologists were on rigs to do their jobs, "Geologists, it should be noted, are not inclined to flit around oil rigs in frilly nighties, or spend their time flirting with welders: they wear overalls and helmets and work 12 hour shifts like everyone else" (Coote 1979, pg. 206). These two women took their case to the Equal Opportunities Commissions, where the oil company eventually settled with them being allowed to do their job.

And though their case was settled at first, they felt that when they were allowed on the rig they were met with hostility. The two women were eventually banned again from oil wells in the ocean. The two female geologists described their experiences after being on the well, whose managers thought their presence was "nonsense," "We met with no adverse reactions, no problems, no accidents. The Atlantic 1 has had female personnel working offshore when drilling for Elf (among other companies). The rig has at least four double cabins containing a W. C. [water closet, also known as a toilet] and a shower…" (Coote 1979, pg. 206).

The article contextualized the problem of women not having opportunities to perform their duties on oil wells in the ocean. It hindered their promotion and the "prospects" of advancement in their filed. The two female geologists have continued to fight for their place on the well, but have not gotten any response, and their inability to perform their job duties on oil rigs have caused shortages with staff at the company, as other male geologists have had to do additional duties. The article implied that though there was a Sex Discrimination Act in law, it did not help the culture that excluded women, as the article went on to name other women scientists who were not allowed to work on platforms in the North Sea from Oil companies like Shell. The article argued that the "accommodations" that were needed were both easy and reasonable, "…a company which spends tens of thousands of pounds a day to keep one vessel afloat could afford a few quid to put up a 'Ladies' sign and replace a shower curtain with a solid door. (She has researched the problem thoroughly, and nothing else seems necessary)" (Coote 1979, pg. 206). The article then asserted that the hesitation to put women on oil rigs might be a British only problem, as the article claimed that an unnamed Norwegian rig had around three hundred female employees on the crew.

Discrimination and sexism on oil platforms continues to be a present problem in the United Kingdom (Saner 2013). Sarah Darnley was tragically killed in a helicopter crash and is one of the first women killed in the North Sea in the search for oil. Though numbers of women are increasing in all aspects of employment in oil and gas drilling for women in the North Sea, women have unequal and inferior facilities on the platform. One engineer, Sophia Kellas-McKenzie, a 28-year-old woman stated, "Some men don't think it's right that women are on the field, but generally most are great and I enjoy being offshore" (Saner 2013). She included her experiences that she is constantly underestimated by the men on the oil rig: "when one or two [men] have told me I couldn't possibly help them with a physical task because I must be too weak, even though I'm probably stronger than a lot of them" (Saner 2013). The article ended with some of the same sentiments that were expressed in glamour profiles of female geologists in the World War II era and postwar where they were trying to acknowledge sexism but played down its significance: "the langue on the right might sometimes be blunt, even aggressive sometimes – "'you can't be sensitive,' she says. But it's not really exist. 'It's not as bad as a lot of people seem to think it would be for a woman'" (Saner 2013).

9.8 Optimism and Adventure: Periodicals and Positive Narratives

Though stories of discrimination and accounts appeared in the newspapers in the 1970s, there was a remaining spirit of optimism for women in geology, and women advancing their educational aspirations. Other stories found in *Life* promoted female only crews in the deep ocean exploring for Oil. In 1970, *Life* portrayed the glamorous life of female scientists working under water with the photospread and article, "A Nest of Naiads: Five Female Scientists Spend Two Weeks Undersea" (Life 1970). All five women were working on graduate research and took turns living down under the sea in the Virgin Islands. The photospread, which appears in Fig. 9.13, mentioned that the only accommodation that they needed at the underwater station was "…to remove a *Playboy* pinup and to add an opaque shower curtain" (Life 1970). The leader of the female scientists asserted that the only difference in abilities down in the deep ocean was that men could "grow beards" (Life 1970).

Other stories ran in newspapers that argued that women were defying expectations and working in the field. In 1976. *The Leader-Post* ran a profile about Mary-Lou Hill entitled "The bush is no longer a barrier" (The Leader-Post 1976). The article framed the career aspiration of Mary-Lou Hill as part of the new opportunities in geology for women that had been hard fought and largely accomplished. It talked about discrimination in geology as a past problem, "There was a time, not so long ago, when women wanting to enter geology faced a granite wall of opposition from within the profession. Those few women who refused to be discouraged and insisted on becoming geologists often found themselves confined to 'office geology' and administrative positions. A female field geologist, was, to use the geological jargon, an 'anomaly'" (The Leader-Post 1976). Women were pursuing geological education and it improved in the last decade in places like Canada. The problems, which the article blamed on companies and government, seemed to be resolved, "The problem was partly the reluctance of private companies and government

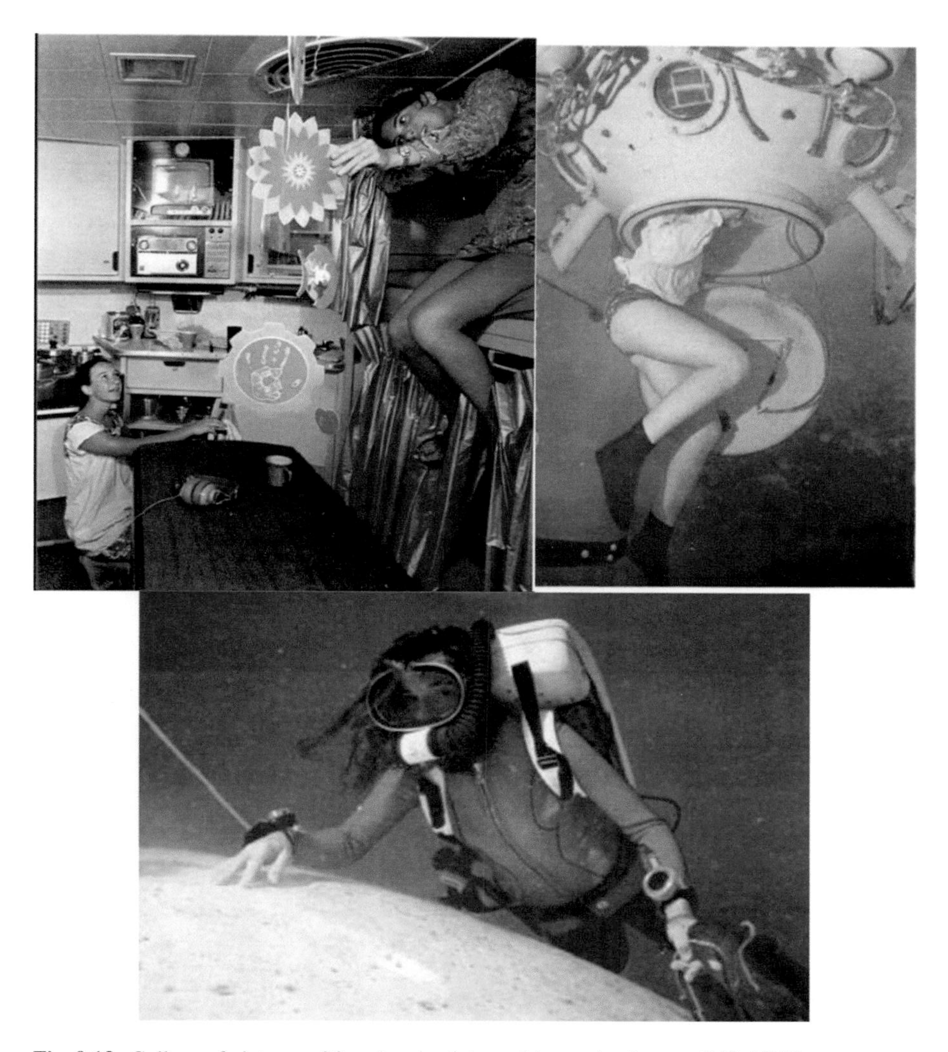

Fig. 9.13 Collage of pictures of female scientists working under the sea. (Life 1970)

departments to hire female students as a part of the summer exploration crews. This was often the type of experience would-be employers were looking for. The result was that women, even those with good marks, rated low when it came to recruiting geology graduates" (The Leader-Post 1976). The article even argued that "taboos against women in the field" had all but "disappeared" (The Leader-Post 1976).

The article promoted that companies were hiring women to work in the field, and most of the women that Hill studied with at the Canadian Carleton University found work, except for those that she claimed did not pursue jobs in sufficient time or interest. Hill also felt supported trying to enter the field, "Ms. Hill's family has generally supported her desires to become a geologist ever since she developed an interest in the subject in high school." But her mother reacted to her employment in field work as questioning "Isn't that kind of butch." Using the term "butch" implied that this was work identified with femininity (The Leader-Post 1976).

The field work was hard, difficult, and dirty, but she felt that most of the men on fieldwork did not really think of her as different, or react at all, "I can't get a serious reaction but I don't think they [men] mind having a girl working with them. As far as working in the bush goes, I don't get any different treatment" (The Leader-Post 1976). She embraced the challenges that were brought by working in the bush and stated that she hoped she was "…doing the same thing[,]" when she was working next summer. The caption to her picture repeated the statement that everyone was the same and that female geologists wanted to be treated similar to men (see Fig. 9.14):

The emphasis on history was also seen as a method to raise morale for women in geology. The magazine *The New Scientist* ran a historical article entitled "Perspective: Ladies with Hammers" in 1979 (Appleby 1979). In acknowledging the renewed popularity of geology, collecting, and fossils, Valerie Appleby argued

Fig. 9.14 Mary-Lou Hill at work in the field. (The Leader-Post 1976)

that the contributions of women in geology have been overlooked and not paid their proper due attention. She framed the changes in geology from the nineteenth century to her own time as a time of radical transitions for women, "The treasure-hunter and collector have given way to the scientist and today, women geologists—too numerous now to name—are extending even further into the new fields of plate tectonics, comparative paleontology and the study of the origin of life itself" (Creese and Creese 1994; Kölbl-Ebert 2020; Appleby 1979). She emphasized that history would inspire women to enter the field and that looking into to the past to find female contributions to geology would help women of today: "If these exciting new advances are helping to inspire a popular revival in the subject today, no one would be happier—and perhaps more surprised—than those first ladies with hammers who struck such blows for geology all those many decades ago" (Appleby 1979).

Newspapers also ran articles that expressed the excitement of students to study geology, and geology related subjects prior to going to college. In 1974 in the article "Geology Comes to Life" both male and female students were expressing interest in studying geology related majors in college because of the problems with energy in the 1970s. The article reported that a geology educator saw optimism for pursuing geology in higher education in both male and female students:

> "His students represent both sexes, and they now come from all sections of the country. A recent survey of major educational institutions indicates most schools of higher learning offer energy-related courses. More and more young people—including women—are indicating they will pursue careers related to exploration and development of natural resources." (Martin 1974)

The article reported that recent, industry studies, found that the immigration and image of the geologist had also changed, "The report wipes out the myth that the geology professor is a man with a pickax on his desk who talks constantly of exotic voyages overseas" (Martin 1974).

Expanding the image of the female geologist was the space program in the United States, the National Aeronautics and Space Administration (NASA), and the space shuttle program of the 1980s. Astronaut Kathryn Sullivan was working with six other women to travel on the Space Shuttle (Lillie 1978; Acqua 1988).[7] Sally Ride, a physicist, became the first American woman in Space and was part of the group of six women (Lillie 1978; Acqua 1988). Aspiring astronauts were encouraged to study science in college, including geology, among other personal and physical skills (Glasgow Herald 1978; Acqua 1988). Excited newspaper articles, including those published internationally, included Sullivan's biography as both the youngest member of the group at age 27 and also a Ph.D. candidate in geology (Lillie 1978). Sullivan was also the first woman from the United States to walk in space in 1984. Other female astronauts at NASA included the mechanical engineer and petroleum engineer N. Jane David, who had received a Ph.D. She had worked

[7] There are several excellent histories of women in the space program and in the history of space exploration. Please see Weitekamp (2004) and Sherr (2014). There are also accessible histories of the space shuttle program in American history, such as Sivolella (2017) and Duggins (2007). There needs to be more literature about the history of women in the space program and in space studies.

as an engineer at NASA, after a two-year career in petroleum engineering at Texaco, and was elevated to astronaut in 1988 (Acqua 1988).

In a break from many of the other female geologists portrayed in this book, Sullivan was critical of the adventure and romantic portrayals of her accomplishments in space. In 1985, an the *Record-Journal* ran the headline that "Woman spacewalker downplays 'fun[.]'" (Record-Journal 1985). Sullivan wanted to emphasize the seriousness and how she perceived the purpose of her work in NASA: "I didn't join NASA to ride Space Mountain (a Disney amusement ride)…I joined NASA to help…go after the answers[.]" (Record-Journal 1985). The article argued that she was not against the adventure of space, but she wanted the public to understand that was not the whole story "The 'golly gee whiz' (of space flight) is an awful lot of fun, but not at the expense of the technical challenges… 'I wish we could communicate better the intense pleasure we get from the technical part of making these dynamic missions work[.]'" (Record-Journal 1985).

She was uncomfortable with the limited attention people got for space missions and did not want people to think everything in space was all glamour. For instance, she said that the public would not find exciting, such as putting information and commands into the computing systems. Sullivan also commented that portrayals of space flight were incomplete and unintelligible to the average person because there were so many small and important parts of mission. She wanted to dismiss the idea of the public that space flight was "'fun and play time'[which] is just as small part of what goes on, she said" (Record-Journal 1985).

Sullivan, however, did not "consider herself a geologist" according to the article. Sullivan explained her perception of her identity as a member of the crew "My profession is mission specialist. Geology is a strong interest but on a very minor scale compared with NASA[.]"(Record-Journal 1985). But she did have an enthusiasm for space travel and women participating in space travel, "There's still a lot of work to be done." The article continued to record how Sullivan saw her contributions, "Sullivan said her spacewalk was 'a minor note' but 'a very happy one because I wanted to do it—not to become the first American woman. I'm not much of a records type person.'" She also emphasized that not every space flight has to be a great feat of historical accomplishment, pointing out to readers that regular flights were acceptable and "doesn't have the significance as a pioneering, courageous thing. There's an element of risk but it's not the same as the Wright Brothers or the very first flight of the very first shuttle" (Record-Journal 1985).

The ideas that Sullivan expressed were a major break from the glamorization of female geologists in this book, and against the type of articles that other newspapers were running that described female astronauts in 1979 as adventurers and used the words that Sullivan wanted to deemphasize in framing meanings of space flights. An example headline included, "Adventurers in Space Flight: Six Extraordinary Women Look to the Stars and See Their Futures."[8] Other women that Sullivan trained with were portrayed in similarly heroic terms, such as articles about Sally

[8] Capitalization was changed in the quote from Hoffman and Tekulaky (1978).

Ride, proclaiming the historical importance of what was occurring in the US Space Flight: "A High-Flying Frontier for American Women" (Pound 1983).[9] Sullivan also deemphasized the need for help from women like Ride, quoted in 1984, when asked whether she needed help from co-crew women like Ride, "'I don't need much coaching,' Sullivan said when asked if Ridge helped her prepare for the mission. 'I just go do my job'" (Gainesville Sun 1984).

9.9 Conclusion

The new glamorous working woman in geology was fighting against conformity in fashion and culture. Books, newspapers, and cultural guides questioned why women wore clothing, hairstyles, or fashions because they were trendy, and advocated that women wear and work the ways that they pleased. The 1970s cultural and social revolution in regard to gender called for women to become more introspective in their roles in relationship, work, and society. The oil industry was caught up in cultural and social changes as well. The oil industry had a culture of racism and sexism, as demonstrated in the intensive and large discrimination lawsuits in the 1990s.

Female geologists also reached out to the African-American community in some cases and taught geology courses in the community. Black publications promoted the working woman geologist in similar glamour terms as postwar profiles of white geologists. Other profiles of female geologists throughout the 1970s and 1980s provided more and more comments and insights about discrimination and challenges that women experienced working in geology and petroleum. Women were also entering management positions. The conversation in profiles of geologists associated glamour involved critical comments of industry or the working experiences of women. The acceleration of the women's movement likely encouraged newspaper profiles of female geologist to include more information about the challenges and struggles of women working in geology.

The women's movement was encouraging of women to share their true and even less than glamorous experiences working in oil and geology. Journalists and publications, and likely oil companies, continued to push a narrative that emphasized the glamour and opportunity of women working in geology by emphasizing its adventurous nature. The increased participation of women in the space industry would produce new and interesting profiles of women working in geology or that had geological training, but these profiles would emphasize the challenges, as well as the success, of women in geology.

---

[9] Capitalization changed in the quotation from Pound (1983).

References

(1950) Women match men in careers, but they must excel in work. The Altus Times-Democrat. 15 November
(1954) YMCA north side presents new face. The Afro American. 2 October
(1963) Professional workers include more women. The Afro American 24 August
(1970) All things being equal..dot hired her! Ocala Star-Banner. 22 July
(1970) Young woman in man's profession: "equality" for geologist. Sarasota Herald-Tribune. 17 July
(1970) A nest of natalids: five female scientists spend two weeks undersea. Life 49(6) 14 September: pp. 50–51
(1971) Sam bucks male bastion in mining field. The Southeast Missourian 4 November
(1973) Bias award is $17,750: woman charged job discrimination. The Toledo Blade. 1 February
(1976) Instead of college she joined a union. Eugene Register-Guard. 13 February
(1976) The bush is no longer a barrier. The Leader-Post 24 July
(1977) Woman geologist aboard drilling vessel. The Evening News. 8 September
(1983) Women's colleges promote their role in education. Lewiston Daily Sun. 1 July
(1984) Shuttle test launch successful. Gainesville Sun 16 September
(1985) Woman spacewalker downplays 'fun'. Record-Journal 9 February
(1997) Oil industry is reeling from sex and race cases. Kingsman Daily Miner. 6 January
Acqua JD (1988) Women astronauts no longer 'primp for orbit'. Altus Times (2 October)
Aig M (1982) Joining the ranks: women learning to survive in work world. The Nevada Daily Mail. 18 July
Alama MR (1966) Hubby fired her: now she's only woman oil executive in country. The Post-Standard. 27 April
Alexander O (2022) Elizabeth Duncan Koontz. Black Past. https://www.blackpast.org/african-american-history/elizabeth-duncan-koontz-1919-1989/
Anderson TH (2004) The pursuit of fairness: a history of affirmative action. University of Oxford Press, New York
Appleby V (1979) Ladies with hammers. New Scientist 29:714–715
Basosi D, Garavini G, Terntin M (2018) Counter-shock. The oil counter-revolution of the 1980s. Bloomsbury, New York
Bass EMB (2020) "That these few girls stand together:" finding women and their communities in the oil and gas industry. Oklahoma State University, Dissertation
Black J (2013) Making the american body: the remarkable sage of the men and women whose feats, feuds, and passions shaped fitness history. University of Nebraska Press, Lincoln
Bloch HD (1965) Discrimination against the negro in employment in New York, 1920–1963. Am J Econ Sociol 24(4):361–382
Brown RL (2020) Where are the black geoscientists? Online campaign calls for diversity. Christ Sci Monit. 11 September. https://www.csmonitor.com/Science/2020/0911/Where-are-the-Black-geoscientists-Online-campaign-calls-for-diversity
Cochrane H (1979) Job problem over women on oil rigs. The Glasgow Herald (4 July)
Coote A (1979) Fear of the frilly nighties why are women barred from Britain's oilrigs? New Statesman 97:206
Creese MRS, Creese TM (1994) British women who contributed to research in the geological sciences in the nineteenth century. Br J Hist Sci 27(1):23–54
Davis H (1976) Women in mining: geologist notes bias. Spokane Daily Chronicle (22 March)
Duggins P (2007) Final countdown: nasa and the end of the space program. The University of Florida Press, Gainsville
Farinde AA, Lewis CW (2012) The underrepresentation of african american female students in stem fields: implications for classroom teachers. US-China Education Review B 4:421–430
Fitzpatrick K (1986) It happened to me: rock bottom. Savvy 7(7):18–19
Friedman D (2022) Let's get physical: how women discovered exercise and reshaped the world. Penguin, New York

Gries RR (2017) Anomalies: pioneering women in petroleum geology, 1914–2017. Jewel Publishing LLC, Lakewood
Hanson S (2009) Swimming against the tide: african american girls and science education. Temple University Press, Philadelphia
Hanson D, O'Donohue W (2010) William whyte's "the organization man": a flawed concept but a prescient narrative. Manag Rev 21(1):95–104
Hoffman J, Tekulaky M (1978) Adventures in space: six extraordinary women look to the stars and see their futures. Rome News-Tribune (19 March)
Jacobs M (2016) Panic at the pump: the energy crisis and the transformation of american politics in the 1970s. Hill and Wang, New York
Kluessendorf J, Mikulic DG (2018) Katherine greacen nelson: advocate for the public apprecia-teion of earth science. In: Johnson BA (ed) Women and geology: who are we, where we have come from, and where we are going? The Geological Society of America, Boulder
Kölbl-Ebert M (2020) Ladies with hammers—exploring a social paradox in early nineteenth-century britain. Geological Society 506:55–62
Kootz ED (1970) Women at work: the women's bureau looks to the future. Mon Labor Rev 93(6):3–9
Laweson D, Conlon J (1971) Beauty is no big deal. Bernard Geis Associates, New York
Lillie H (1978) Women who will reach for the stars. The Glasgow-Herald (13 February)
Link M (1973) Women engineers-demand exceeds supply. The Argus-Press (23 February)
Lyons RD (1972) Female incomegap grows. The Sunday News Journal (28 December)
Martin FT (1974) Geology comes to life. Beaver Country Times (30 October)
Matelski EM (2017) Reducing bodies: mass culture and the female figure in postwar america. Routledge, New York
McCandless AT (1992) The higher education of black women in the contemporary south. The Mississippi Quarterly 45(4):453–465
Merdecal MD (1972) Ex-harlanite really digs rocks. Harland Daily Enterprise. 31 May. Originally published in the News and Observer newspaper in Raleigh, North Carolina. The original author is attributed with credit for the piece
Myers GC (1960) Few girls enter field of science. The Windsor Star (7 April)
Paull RK, Paull RA (1984) Memorial to katherine greacen nelson: 1913–1982. https://rock.geosociety.org/net/documents/gsa/memorials/v15/Nelson-KG.pdf
Pound L (1983) A high-flying frontier for american women. Spokane Chronicle (13 June)
Roth M (1984) Program guides women into science. The Michigan Daily 95(10) September 16
Rubio PF (2001) A history of affirmative action, 1619–2000. University of Mississippi Press, Oxford
Saner E (2013) Women on oil rights: 'some men don't think it's right that women are there'. The Guardian 27 August: https://www.theguardian.com/money/shortcuts/2013/aug/27/women-oil-rigs-helicopter-deaths
Schneiderman JS (1994) Curriculum transformation in the arth sciences: women's studies and geology. Transformation 5(1):44–56
Seiler C (2008) Republic of drivers: a cultural history of automobility in america. University of Chicago Press, Chicago
Sherr L (2014) Sally ride: america's first woman in space. Simon & Schuster, New York
Sivolella D (2017) The space shuttle program: technologies and accomplishments. Springer, New York
Slater RB (1994) The blacks who first entered the world of white higher education. J Blacks High Educ 4:47–56
Tolley K (2003) The science education of american girls: a historical perspective. Routledge, New York
Travis D (1956) Milwaukee mirror. The Afro-American. 7 April
Urofsky MI (2020) The affirmative action puzzle: a living history from reconstruction to today. Pantheon Books, New York
Wdowik M (2017) The long, strange history of dieting fads. The Conversation 6 November

Weitekamp MA (2004) Right stuff, wrong sex: america's first women in space program. Johns Hopkins University Press, Baltimore
White III, Frank (1984) Lady pioneer in the oil field. Ebony. January:60–63
Whyte WH Jr (1956) The organization man. Simon and Schuster, New York

Chapter 10
Glamour Returns: Images, Heroics, Dolls, but Far from an Ending

10.1 Far from an Ending

In 1982, women in geology and petroleum geology were continually categorized and praised as being "unusual." Jeri Cooley, profiled in the Montana-based *The Bryan Times*, was working in the field as a petroleum geologist. The paper noted her experiences as "Being a geologist in an oil field is an unusual occupation in itself, particularly for someone from this area and especially for a girl" (Geesey 1982). She was a geologist who understood how to measure hydrocarbons in potential oil wells that, on her judgement, would result in the well being tapped for oil.

Like many of the women portrayed in the book, she became interested in geology in college. Cooley graduated from Muskingum College, and started out as a French major, but changed her plan of study after taking two mandatory science classes, which she enjoyed. She recounted her positive experiences in colleges and how she changed her emphasis to geology, "She planned on studying geology as a hobby, but found she couldn't take her next French course until her junior year. By the end of a sophomore year of studying geology, she decided to switch majors" (Geesey 1982). She also noted that the "economics" of the matter moved her, as she perceived the job prospects for geology would produce more job opportunities than her French major.

Cooley liked her job, enjoying the travel and economic rewards, and was thinking of moving more to a consultant geologist position, and continue to improve at her job. She noted "exciting" parts of her job, such as gas flare that occurred, which Cooley explained to her mother in a letter, and was also thankful that she did not perish, "'We were certain we'd go up in a big mushroom cloud,' she wrote" (Geesey 1982). She included more information about the emergency, where the roughnecks, or oil workers, tried to make the female geologists working on the well leave, even though Cooley was in a position of safety,

E. A. Driggers, *Glamour and Geology*,
https://doi.org/10.1007/978-3-031-64525-9_10

Jeri Cooley

> All of a sudden, a roughneck burst through the door. She said that we “beautiful dolls” should either “vacate or incinerate” [sic] there had been a big explosion on the rig floor, but we’d not heard it above the engines. “He went back up to the floor,” the letter said, “and promptly passed out from the deadly gas. The toolpusher (rig supervisor) had gone to call for an ambulance, so Cindy (another well log analyst) and I were the only ones on location still OK.” (Geesey 1982)

Cooley’s expertise in science likely saved her life, and helped her remain one of the few workers at the well, conscious so that she could help. Crewmembers had been injured in the initial explosion, so she had to rescue them all, “There had been a terrible accident on the floor. The driller had broken his hip in the explosion, most of the crew was either unconscious or about so. [sic] Cindy and I put our oxygen tanks and rescued those crew members” (Geesey 1982).

In her letter to her mother, Cooley expressed disbelief that she and Cindy were able to not pass out themselves, as the tanks only held around thirty minutes of oxygen. There were a lot of workers “hands” that they had to give oxygen in order to get them to return to breathing. Her letter continued to vividly describe the dangerous situation: “It was such a dangerous situation with all of the mud shooting out of the hole, the gas everywhere and sparks. The whole rig would have gone up in flames!” (Geesey 1982). Cooley praised her safety courses, as she had to act in an

Fig. 10.1 Cooley and the gas explosion (Geesey 1982)

urgent manner, and even do so without time to think. They were thanked for their quick actions by many of the workers. Cooley and Cindy won the support and admiration of their male co-workers: "'They'd all confessed they'd not thought much of women in the patch, but were so thankful we had been there that night!!' the letter said" (Geesey 1982).

These accidents can occur with specific geology around the well, but this was noted by her as a very exciting part of the job. And working in oil was different than most people could even imagine, and the article punched up that language, "She said despite all the danger and excitement, she loves her job. 'The oil patch is a totally different world from reality. You meet all kinds of people'" (Geesey 1982). And she would not trade her job for another, nor would she have thought to change her major: "Although she said she would like to return to school, she wouldn't pick up her studies of French. She would stay with geology" (Geesey 1982).

The romantic image of geologist working in petroleum was advertised to children in the 1980s. In the children's show *3-2-1 Contact*, in an episode, a female clothing designer who was also a project engineer was featured who designed the space suits for NASA. The episode featured several female scientists who were portrayed as "role models" and that no matter what their clothing, they were doing a diverse set of scientific work: "As role models, there women are living proof that scientists are likely to be wearing hiking boots or space suits as a white laboratory coat—and that nowadays the person wearing the boots or space suit may well be a woman[.]" (Gadsden Times 1980).

Bonnie Robinson, who was 27 years old, was a guest on the show that was explaining her job working as a geologist in petroleum geologist. She explained to the children hosts of the show that her job was similar "to being a detective" (Gadsden Times 1980). Robinson explained that her search for oil required her to essentially read the rocks, "I search for clues in rock formations that might indicate there is oil underneath[.]" (Gadsden Times 1980). Her interests, like many of the women profiled in the book, started with simply collecting rocks in her childhood, when she lived in Washington, D. C. Though she lived in an urban city, she found rocks in local parks, "I became more and more interested, used to haunt the city's parks to add to my specimens" (Gadsden Times 1980). She interjected that men in her life were not supportive and tried to humiliate her for collecting rocks, "I had to take teasing from guys because there were so few female geologists the[n?] [.?] We're still a minority, but the field is wide open for women. It's interesting work and provides constant opportunity for travel" (Gadsden Times 1980). Though women were not a large part of petroleum geologists, Robinson advertised the profession as one that women can pursue and one of adventure. The astronaut and geologist Kathy Sullivan also appeared in that same episode, which worked to punctuate that idea.

Sullivan was asked how she became an astronaut and tried to add a bit of humor, responding that she replied to a job ad. But her tone changed to portray NASA as a space open for women to pursue employment: "Then she adds seriously that NASA advertises periodically for scientists who are interested in the space program and they need people from a wide range of fields" (Gadsden Times 1980). As a geologist, she really hoped that she would be able to examine the rock formation on Mars. She included her future hopes, but also explained to the children hosting the show how she became interested in geology, and her interests was heightened through a really positive educational experience. Sullivan recalled that, "'When my interest in science was first piqued by a college professor who showed me a grain of sand under a microscope [.?] I never dreamed that my interest in the earth would send me into outer spa[ce?]' she says[.]" (Gadsden Times 1980).

The description of the show punctuates the importance of the accessibility of science to women and empathized the idea that women can have meaningful careers in science. The article, including an image that is featured in Fig. 10.2, put the episode in important terms, "Girls need to know t[that?] careers in science are ope[n?] [to] them and they are not dull, that [in?] fact they can be among [the?] most exciting jobs of all[.?] Meeting these women should give them the message [.]"[1] (The Lewiston Journal 1980).

[1] The first source had a lot of the print copy faded, so I looked up another version of the article, which also included a photograph of Sullivan. The photograph and additional quotations came from (The Lewiston Journal 1980).

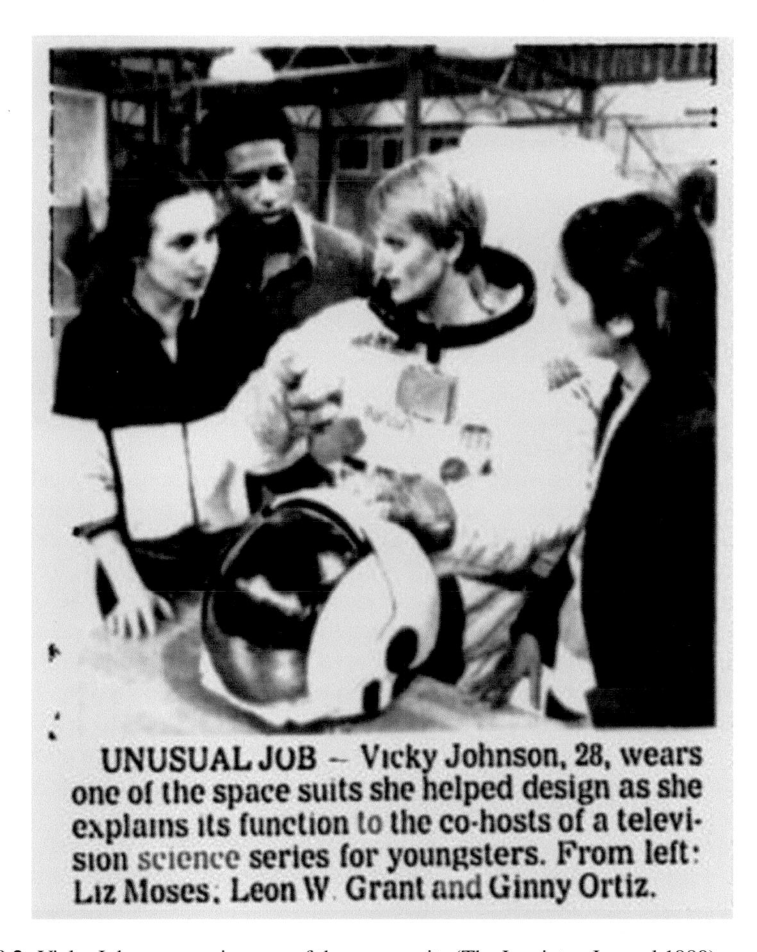

Fig. 10.2 Vicky Johnson wearing one of the space suits (The Lewiston Journal 1980)

This same language that appeared on televisions in 1980, appeared in literature promoting science to women and minorities in 2005. In the October–November 2005 issue of *Women of Color*, a write-up of the value of tech camps, put on by International Business Machines (IBM), continued to empathize that science was an exciting profession. But analysis of the gender of participants in mathematics finds that women depart from technology because of the perception that women lose interest, as the language becomes defensive that sciences and technology was perceived as "dull" (Women of Color 2005). The magazine discusses the technology camps and spells out their purpose, as they were to change expectations of the children attending them, "Studies show that young girls enjoy math as much as boys, but by the eighth grade, twice as many boys show an interest in pursuing careers in science, engineering, and math. We've got to make young girls understand that a career in technology does not have to be dull or boring. It's just the opposite.

Technology and science related careers offer opportunities to be creative, to become a leader, to give back to your community, and to establish financial independence" (Women of Color 2005).

Petroleum geologist and geologists in general appeared in the career series of the popular toy *Barbie*. There were several editions of *Paleontologist Barbie*, including a 1996 release of an African American *Paleontologist Barbie*. A special edition was released that included many accessories and encouraged purchases of *Barbie* to play at digging up dinosaur bones. The child added water and sand in the performative to recreate the excitement of discovering fossils. The 1996 editions of the Caucasian and African American *Barbie* had more accessories including a rock hammer, specimen bag, dinosaurs, canteen, and other dinosaur memorabilia. The shirt is colorful and included dinosaurs, and *Barbie* has footwear, but I cannot determine what type of footwear. The series had a much more robust name, the "Career Collection: We Can do Anything."[2]

In 2011, *Barbie* was released again in the series *Barbie: I can Be* as a paleontologist, who included more active accessories such as a shovel. *Barbie* is dressed in a functional wardrobe that seems functional for the field as well. Her dress accessories included high lace up boots suited for field work, as well as trowel for working in the dirt to hunt for fossils. Her hat and jacket are fashionable and have brightly colored patterns and colors. She has an undershirt with a dinosaur. Her pants contain pockets and were in the "cargo" style of the early 2000s, but were adjustable culottes for potentially hot weather. Her jacket also had brightly colored pipping. Her other accessories include fossils that she discovered and they are displayed in a dirt sifter. There was an additional puzzle and access to online content that would allow for the child to further pursue information on being a working paleontologist. The 2011 doll received an endorsement from the Smithsonian.

Later on, there were co-branded *Jurassic Park* outfits for *Barbie* as well. *Jurassic Park* was a highly profitable movie franchise that involved science and the recreation of dinosaurs. The franchise included several female scientists as main characters. Neither of the *Barbies* were produced in the body inclusive "fashionista" series, which was based in body positivity. The 1996 and 2011 *Barbies* were produced on the traditional *Barbie* bodies. Newer series, which featured female scientists, such as the line of *Barbies* was produced during the International Women's Day in 2023. The series features different type of skin tones and body styles to represent the real female scientists that Mattell based the series on. Other, newer, female science-based *Barbies* included different types of hair colors and accessories that represent newer fashion trends and ways that women were working in science.

Competitors to career-based *Barbie* have appeared in other toy series. The famous Lottie Doll was released as a fossil hunter. Lottie comes with field work boots, shorts, and a jacket, with fossils and a magnifying glass. Lottie also has a nautilus shirt and several nataoloid fossils. The doll also comes accompanied with biographical information on famous female geologists, such as cards featuring Mary Anning. The Lottie dolls are dolls that encourage female children to study

[2] I found several examples of this toy through Amazon and Ebay.

science and STEM disciplines. The dolls were also designed to be different from the hyper feminized *Barbie* dolls and were a response to hyper-sexualization of female-based toys (Thomas 2003; Rogers 1999; Lord 2004).

News stories about geology and petroleum education advertised the openness of industry to women in the same manner that similar stories discussed educational opportunities for women during World War II (Lindquist 1961). In February of 2020 in the story "WHS Petroleum Class is Proving Positive for Oil Industry," high school students in North Dakota had the opportunity to take a "Petroleum Career Class" with a petroleum geologist Gerald McGillivray, who had worked in the industry for around forty years. McGillivray noted that he gets more female than male students, as he challenged their expectations about the petroleum industry,

> I get more girls because they think you've got to work on a rig, or you got to work outside, and I say no there's a lot of technical things. You see a lot more women engineers, and even the outside jobs. See a lot of women truck drivers. It's not a man's world. It used to be, maybe, but it's gotten a lot better for women…[.] (Tanner 2020)

The message seemed to resonate with female students, as the article reported that more women in high school were seriously considering careers in the oil industry. One student interviewed, Diana Gutierrez, commented that she was surprised about the petroleum industry: "At first I kind of didn't, until I got into this class. I started looking into it, and I think petroleum engineering would be a choice for me" (Tanner 2020).

Ideas that romanticize the image of the female petroleum geologists remain in publications, such as *Oil Woman*. Many of the profiles in this magazine look very much like the write ups of June King in the 1940s. The magazine has the secondary title that it was "The Magazine for Leaders in American Energy." The editor in chief letter that was included in the first edition promised to profile "the most incredible women in the industry" (Ponton 2021). Rebecca Ponton, the Editor in Chief, linked her magazine with more access for women into leadership, discussing the election of Kamala Harris to the vice-presidency of the United States. Ponton noted that the first cover portrayed a woman of Asian Heritage, Donna Cole, who was a chemist and CEO of Cole Chemical. She also noted that Cole was portrayed on the cover in honor of the month of May, as it was Asian-American and Pacific Islanders Month, but that all women have a contribution to the careers and history of women, "We may not have a world stage or a theater stage, but we all have a sphere of influence. We all can mentor—even unknowingly—through our words and deeds; you never know whose life you might touch" (Ponton 2021).

Profiles and stories in the magazine work to advocate to female readers how to enter into the larger energy field that includes resources from petroleum, coal, and alternative energy sources. Profiles of industrial leaders use many of the strategies of writing about women that have been discussed in this book. The section "Woman on Board" that profiled Arcilia Acosta used a lot of the same symbols and image construction of women like June King in World War II. Acosta is pictured with a hard hat and described as having a demanding schedule. She had companies that are involved in searching for oil in the Permian Basin, and she has major "clients" such as Chevron.

The image of Acosta emphasized someone who was dedicated to her family, as she emphasized working hard to support her sons. She was also portrayed as a passionate risk taker, who left a stable career in marketing for the energy field. She was waking up at 3 a.m., working at her startup company, and then returning at 7 a.m. to drive her kids to school. The profile empathized her life philosophy as "*have no regrets*" (Jackson 2021). Acosta was a new type of "Oil Woman" who also acknowledged the gender issues in her field, and was portrayed as a woman who was rare in a mostly male dominated field. But these profiles, like the profiles from World War II, embraced some of the challenges of being a working woman. The profiles of 2023 delved deeper into the challenges and were bolder in their descriptions of the challenges to women in the energy industry. "In building her business, Acosta learned to *work through struggles*. When she joined the engineering and energy industries, she didn't realize how few women were in those fields. Unfortunately, she admits she still can count on one hand the number of women at the job site or in the room. Acosta faced her own battles with sexism and racism" (Jackson 2021).

She shared such a story of racism and sexism that she personally experienced, "She relates the story of her subcontractor who walked up to her on her job site in flip flops and shorts and said, 'I hear you're the Latino woman who's winning all the jobs'" (Jackson 2021). Acosta responded by making the worker leave the job site because of his lack of professionalism, "Instead of overtly addressing the microaggression, Acosta responded with a brilliant and honest approach that made her point about who's the leader and in charge: 'And who might you be? And could you please leave my site and come back tomorrow in proper safety gear?'" (Jackson 2021). She argued for boundaries and values, but inherent in these new profiles of women working in petroleum geology and the energy industry in general is the emphasis on supporting other women. Acosta had a personal value that she made her motto: "Empowered women…empower women" (Jackson 2021). She also advocated for self-confidence and the embrace of intuition for working women, "She encourages us all [women] to 'trust our vibes, our gut instincts, and the red flags that something is not right; we women are very powerful with our institution'" (Jackson 2021). If women follow these ideas and suggestions, women will be better represented in leadership positions in industry, such as serving on corporate boards.

Oilwoman has a positive approach to women's employment, sharing stories that help women with everything from Human Resources to constructing a useful social media account. One story in June of 2023 that ran was an article discussing how women can work together to traverse challenges in the energy field, such as discrimination or how to react to unseemly behavior by co-workers (D'Eramo 2023). The short piece advocated the idea that women could be "allies" to each other in the workplace. The article openly acknowledges the problems of patriarchy, or as the included Meriam-Webster dictionary as "A system of society or government in which men hold the power and women are largely excluded from it" (D'Eramo 2023). The article also advocated for women to share their stories as a way of working through challenges that women faced in their careers and workforce. The article explored the ways in which women could challenge male power but working women should not be burden with fighting the whole of society's problems, "It is not women's responsibility to dismantle patriarchal power structures, but we can understand our options to disrupt patterns that create barriers to equality" (D'Eramo 2023).

The author, an employment consultant, Erica D'Eramo, discussed the ideas of sharing stories to educate future generations of women:

> The benefits are priceless. Envision a world in which we consistently find a joy in each other's success, where we can rely on professional sisterhood for allyship and encouragement, where we pass down knowledge and wisdom between generations, and where re reject outdated stereotypes and biases. (D'Eramo 2023)

The article continued to press the reader to try to then turn the narrative, and understand the position of why co-workers were affecting them,

> Self-awareness gives us a priceless moment between reaction and response to select the most effective path toward our goals. Next, we put ourselves in our counterparts's shoes and try to understand what motivations, threats or biases might be influencing them. What knowledge, capability and resources are they operating with? What do we know a fact, and what information is missing? Lastly, we look at context. What are the cultural and structural elements at play? How do history, power dynamics and economic forces influence this situation? Are we break down some of the challenges that lead to unsupportive behaviors and explore how to disrupt these patterns...(D'Eramo 2023)

The article reminded the reader that biases came from within and argued that women needed to "Break the cycle." And that by advocating for other women, even in situation where it is not recognized: "Be the woman who raises up other women when they're not in the room, who has their backs, and challenges sexist language about their peers" (D'Eramo 2023). Other "disruptions" that are listed are technologies that make language more inclusive or "internalized misogyny" (D'Eramo 2023).

Profiles of female geologist in the 2020s continued to emphasize the struggle of women working in geology, and emphasized their triumphs. Bowen, an African-American geologist, had her profile and interview appear in the publication *Shoutout Atlanta*. Ashley Nicole Bowen described her educational background as pursuing a bachelor's degree in geology at the University of Alabama, located in Tuscaloosa, Alabama (Shoutout Atlanta 2021). She went on to work at the Geological Survey for the state of Alabama. Bowen discussed her experiences in schools prior to college, being one of the first women in her family to pursue a science education. Bowen elaborated on her experiences as a black woman pursuing her geology education: "Being first generational in my family and coming from underperforming schools, there was never a mention of the natural science. So after learning the complexities of the subject, I became more and more intrigued and was excited to work as a Petroleum Geologist" (Shoutout Atlanta 2021). Bowen emphasized that though she was interested in her job, she wanted to change her career trajectory, which is an interesting pivot from the profiles that started the book. Bowen explained that,

> It was my team['s] duty to study and find economically producible oil and gas within the state. However, after seven years in the field, I realized that something was missing. I knew that I wasn't meant to sit behind a desk for the rest of my life. So in 2010 I began my journey hosting events. By 2015, I opened my first storefront as an Events Management Company. Specializing in Weddings, Private Events, Corporate Events, linen and Chiavari chair rentals. After grossing six figures in my first six months, I knew then entrepreneurship was where I belong. (Shoutout Atlanta 2021)

Women, like Bowen, felt confident and free to choose their own career path, even if it was pivoting away from geology.

Contemporary accounts of women in geology and oil also ran similar profiles to those of June King in World War II. In 2023, Emma McConville was profiled in the *New York Times*. She had been hired as a geologist for Exxon in 2017 and was working in petroleum geology. McConville was working at a really important oil development project in Guyana. But the profile diverged from the successful profiles in World War II and discussed the problems of the oil industry, such as the drop in prices in oil trading that occurred during the COVID pandemic. McConville discussed being "laid off" during an online call and felt a feeling of extreme "shock." However, she was able to get another job at another energy company that pursued renewable sources like geothermal energy in around four months (Krauss 2023). And she commented that her employment was better, she was managing more projects, and making more money. McConville's profile is likely more of a transition point in the oil industry. McConville contextualized her experiences for the readers, "'Covid allowed me to pivot,' she said. 'Covid was an impetus for renewables, not just for me but for many of my colleagues" (Krauss 2023).

The article went on to discuss how the pandemic caused massive layoffs in the oil industry to around 160,000 workers. Though the oil industry had hired fewer workers in 2023, around 700,000 less in comparison to the last "six years," which Clifford Krauss, the author of the article, estimated to be a loss of 20 percent, workers seem to be ready to transfer their skills from oil to renewable energy (Krauss 2023). Krauss contextualized that workers making that transition would find employment for their skills: "Many workers, including electricians, offshore construction engineers, information technology specialists and environmental surveyors, say the skills they honed in their oil and gas jobs have translated well to the work they are doing now" (Krauss 2023). A conception of the new type of glamour of working woman geologist involves a career that is couragous in translating new skills and previous experiences to new jobs, despite a difficult economy.

10.2 Crafting Their Own Glamour

Though I do not know what the future looks like for women working in petroleum geology, I do expect that the next generation of women geologist being profiled by industry, appearing in newspapers, websites, or even social media, will find use for some of the strategies portrayed in this book, as the sharing of stories and experiences of working women resonates with improving the field and improving the experiences of women in geology. I do hope that the historical emphasis on physical attributes, weight, and pressure to adhere to the male gaze will ultimately disappear from profiles of female geologist. In its place will be the more glamorous ideas of women supporting women, co-workers of all identities reflecting on bias, the striving for better workplaces, and women having the ability to craft their own narratives. Women in charge of their own narratives are a bright future for everyone.

Many of the women that were written about in this book were not well-known figures like Rosalind Franklin or Madam Curie, but without scientifically minded historians, or historical minded scientists, those women would not be well known either. Average women, like those stories presented about June King, reveal even more intriguing and interesting narratives in science. There can be extraordinary stories of genius, but also equally interesting stories of women working on their jobs. Historical figures like June King (Fig. 10.3) were doing the everyday hard work of field geology, as well as working in the laboratory. Unfortunately, as of this moment, the magazine entry in *The American Weekly* that appeared in 1945, is all we have regarding information about June King's life. Recounting the experiences of so many everyday women do we change our perception from the anomalies narrative that was put forth into the newspapers, and instead see the glamour that King and others were beginning to create. Their glamour was slowly expressing their own experiences, frustrations, discriminations, joys, and setbacks working in geology. The most interesting narratives are those that women wanted to create themselves. The first part of the book showed how women embraced the glamorous and ultra-feminine narrative of the working women geologist who did it all: great job, family, war heroes, patriots, supportive wives, and an adventurous work. However, women, with changes in politics and culture, gradually pushed against those narratives put upon them, and created their own, new type of glamour, that they were able to define themselves and what is important in profiling a female geologist.

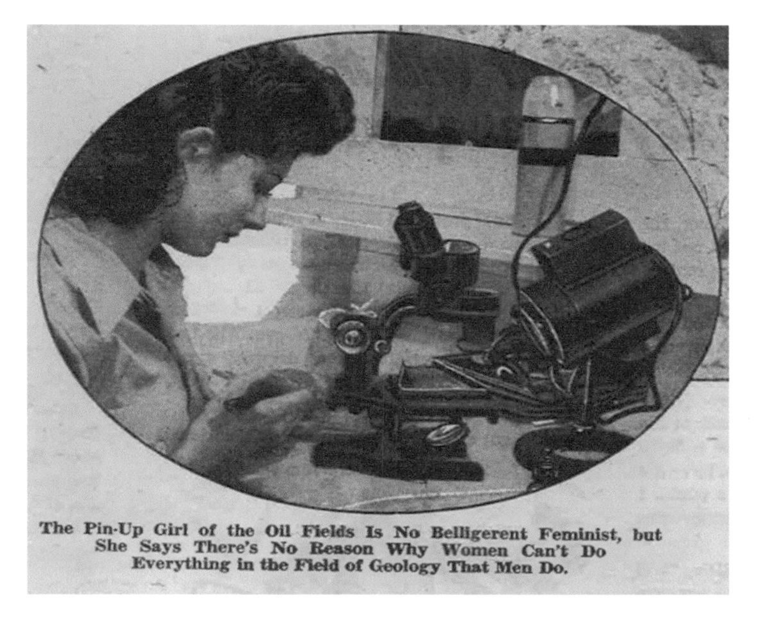

Fig. 10.3 Return to June King (The American Weekly 1945)

References

(1945) Mixing glamor and oil. The American Weekly, September 30
(1980) Who says science is a dull career. The Lewiston Journal, 12 April
(1980) Her clothing designs won't be found in stores. Gadsden Times, 26 March
(2005) Trends. Women of Color. October–November, p 27
(2021) Meet ashley nicole bowen: entrepreneur, founder, woman enthusiast. Southout Atlanta, 15 December: https://shoutoutatlanta.com/meet-ashley-nicole-bowen-entrepreneur-founder-woman-enthusiast/
D'Eramo E (2023) Women as allies—part ii. Oilwoman Magazine, March–April. https://digital.oilwomanmagazine.com/oilwomanmagazinemarchapril2023#page=39. Accessed 18 Mar 2024
Geesey C (1982) Unusual occupation sometimes brings danger. The Bryan Times, 18th March
Jackson L (2021) Arcilia Acosta: a great connector. September-October. https://digital.oilwomanmagazine.com/septemberoctober2021#page=15. Accessed 18 Mar 2024
Krauss C (2023) As oil companies stay lean, workers move to renewable energy. New York Times, 27 February. https://www.nytimes.com/2023/02/27/business/energy-environment/oil-gas-renewable-energy-jobs.html. Accessed 18 Mar 2024
Lindquist CB (1961) Geology degrees during the decade of the fifties. May–June 5(8):15–17
Lord MG (2004) Forever barbie: the unauthorized biography of a real doll. Walker & Company, New York
Ponton R (2021) Letter from the editor-in-chief. Oilwoman. May–June
Rogers MF (1999) Barbie Culture. Sage, London
Tanner A (2020) WHS petroleum class is proving positive for oil industry. https://www.kfyrtv.com/content/news/WHS-Petroleum-Class-is-proving-positive-for-oil-industry-567817731.html. Accessed 18 Mar 2024
Thomas JB (2003) Naked barbies, warrior joes, and other forms of visible gender. University of Illinois Press, Urbana

Index

E. A. Driggers, *Glamour and Geology*,
https://doi.org/10.1007/978-3-031-64525-9